ENVIRONMENTAL POLITICAL THOUGHT

ENVIRONMENTAL POLITICAL THOUGHT

INTERESTS, VALUES AND INCLUSION

ROBERT GARNER

BLOOMSBURY ACADEMIC
LONDON • NEW YORK • OXFORD • NEW DELHI • SYDNEY

BLOOMSBURY ACADEMIC
Bloomsbury Publishing Plc
50 Bedford Square, London, WC1B 3DP, UK
1385 Broadway, New York, NY 10018, USA
29 Earlsfort Terrace, Dublin 2, Ireland

BLOOMSBURY, BLOOMSBURY ACADEMIC and the Diana logo
are trademarks of Bloomsbury Publishing Plc

First published by Red Globe Press 2019
Reprinted by Bloomsbury Academic 2023

A catalogue record for this book is available from the British Library.

A catalog record for this book is available from the Library of Congress.

ISBN: HB: 978-1-1376-0740-9
PB: 978-1-1376-0739-3
ePDF: 978-1-1376-0741-6
ePUB: 978-1-3503-1184-8

To find out more about our authors and books visit
www.bloomsbury.com and sign up for our newsletters.

CONTENTS

LIST OF BOXES

PREFACE

This book is a significant departure from my previous text, *Environmental Politics*, published by Harvester Wheatsheaf in 1996 and Palgrave in 2000, and 2011. That book offered general coverage of the environmental politics debate, covering governance issues at the national and supranational level as well as Green political thought and environmental ethics. This book, by contrast, offers narrower coverage, centring on ideas rather than environmental institutions and processes.

This narrower coverage is justified on two main grounds. Firstly, the literature on environmental politics has expanded to such an extent that it is increasingly difficult now to provide anything other than a very shallow account of it in a book that remains a reasonable length. Secondly, it is rare for University courses (most of which are now only one semester long) to cover all of the content of a general text on environmental politics. Rather, they specialise in government, international relations or political theory aspects of the environmental debate. This book is not merely a volume of normative political thinking on the environment (or Green political thought, as it is usually known). Rather, it also examines empirical accounts of the rise of environmentalism and the political economy of the environment as well as providing extensive coverage of environmental ethics, in general, and animal ethics in particular.

I would like to thank Lloyd Langman for commissioning the project, and Tuur Driesser for chasing me up when I missed deadlines (a regular occurrence!). Thanks should also go to students on my final year Environmental Politics module at Leicester over the years who have enabled me to hone the content of this book, along with its predecessors. As always, responsibility for any omissions and mistakes remains, of course, with me.

Robert Garner
Leicester, October 2018.

1 INTRODUCTION: THE IDEA OF ENVIRONMENTAL POLITICS

This book is about the political ideas current in thinking about the environment. It operates with a broad definition of 'ideas', including normative and empirical dimensions, and a broad definition of the 'political', focusing not only on the state, but also civil society and the individual. Above all, the aim is to explain how a 'politics' of the environment can be distinguished from the coverage of other disciplines and, in particular, from environmental science. To do this, it is necessary to explore, in this introductory chapter, the idea of politics itself.

What distinguishes a *politics* of the environment is the recognition that there are a variety of competing interests and values that need to be identified and explored. Environmental degradation does not affect all currently living humans in the same way. Moreover, the extent to which the goals of the Green movement are embraced depends upon the values adopted and, more specifically, what is regarded as an appropriate ethical position about the relationship between the human and non-human realms. A central component of environmental political thinking is a debate about how far moral entitlements end with currently living humans. Many political thinkers, and not just those with a commitment to environmentalism, hold that we have at least some moral obligations to future generations. Radical Green thinkers want to go even further by suggesting that these obligations extend beyond the human to non-human animals and even non-sentient parts of nature. As a result, future generations and non-humans ought to be included as beneficiaries of decisions currently living humans make, and perhaps also included, too, as members of the polity or political community. The three themes – interests, values and inclusion – therefore form the unifying thread to the chapters in this book.

The rise of the environment as a political issue

Before the component parts of environmental political thought are unpacked further, it is useful to explore the social and historical context in which they have emerged. It is only since the 1970s that the environment has become a salient political issue, and only since the latter half of the 1980s that it has become a mainstream one. Consequently, the study of environmental politics has discarded its Cinderella status. Indeed, by the late 1980s, we seemed to be entering a new Green era, where environmental concern had become the height of fashion. In the developed world at least, opinion polls revealed mounting public concern for the state of the environment; governments responded with legislative and administrative action, consumers demanded environmentally friendly products and producers, with varying degrees of honesty, sought to provide them; recycling centres and bottle banks flourished. Since the late 1980s, the environment has slipped down the issue agenda a little, overtaken by dramatic political and economic events. It is now established, however, as a permanently important feature of political and academic discourse. Sovereign states are now locked into a supranational structure of institutions and processes, initially set in train by the United Nations Conference on the Human Environment convened in Stockholm in 1972 and built upon at the United Nations Conference on Environment and Development (UNCED) held at Rio 20 years later, and at the World Summit on Sustainable Development (Rio+10) held in Johannesburg in 2002. These conferences, together with the series held in pursuit of a solution to climate change, guarantee the continued participation of nations in the search for acceptable solutions to environmental problems.

An obvious initial question to ask is why this idea of environmental protection has become so politically salient in recent decades. The obvious answer is that the environment has become an important political issue because of the deleterious effects of human activity on the planet. Chapter 2 seeks to outline the nature and consequences of this activity. Information on environmental degradation is readily available, and one would be forgiven for concluding that the rise of the environment as an important political issue is related to lay observations and scientific evidence concerning such 'invisible' phenomena as climate change and ozone depletion. Indeed, one factor which ostensibly distinguishes the environment from many other issues is the extent of objective measurement involved. Thus, there is a crucially important technical core to the study of the environment, providing a key role for engineers, scientists and technicians (Weale, 1992: 10). Put simply, if there is a hole in the ozone layer or a build-up of carbon dioxide (CO_2) emissions, and this threatens the stability of the global environment, then we need to do something about it. And if something needs to be done, then we need to decide what is causing it and do something about preventing it.

The growing sense of an objective environmental crisis, therefore, is an obvious reason for heightened concern. One can point here to the well-publicised environmental disasters of the past 40 years or so – the mercury poisoning at Minamata Bay in Japan in 1959; the slag heap slip at Aberfan in Wales which buried a school, with great loss of life, in 1966; the oil pollution caused by the stricken tankers *Torrey Canyon* (in 1967) and *Exxon Valdez* (in 1988) and the major oil leak in the Gulf of Mexico in 2010; industrial accidents at Bhopal in India, which killed over 3,000 people and injured many hundreds of thousands, and at Seveso in Italy; the near-thing at the Three Mile Island nuclear facility in the USA in the late 1970s and the real-thing at Chernobyl in the 1980s – to name but a few. One can point, too, to important books, conferences and scientific research – Rachel Carson's best-selling *Silent Spring* (1962) which documented the effects on the countryside of pesticide use; the first pictures of the Earth taken from space in 1967 which emphasised the fragile and insignificant nature of the planet and its occupants; the *Limits to Growth* report (Meadows et al., 1972); the previously mentioned Stockholm conference (proceedings published as Ward and Dubos, 1972); the Brundtland Report (World Commission on Environment and Development, 1987). Finally, since the 1980s, these problems have been submerged by the emergence of genuinely global issues such as climate change and ozone depletion.

This objective evidence indicates not only that environmental problems have increased quantitatively – in the sense that the number of environmentally damaging incidents has risen markedly – but that there has also been a qualitative shift. Environmental problems are no longer perceived merely as localised concerns affecting relatively few people and having few long-term consequences. Instead, the central concern has become nothing less than human survival on a planet which, it is now recognised, cannot continue indefinitely to cope with the consumption of non-renewable resources or the absorption of waste products from industrial processes at the levels which it is presently asked to do. The following information, provided in the early 1990s, tells its own story:

> Since 1900, the world's population has multiplied more than three times. Its economy has grown twentyfold. The consumption of fossil fuels has grown by a factor of 30, and industrial

> production by a factor of 50. Most of that growth, about four-fifths of it, occurred since 1950. Much of it is unsustainable (MacNeil et al., 1991: 3).

Since then, the situation has deteriorated further. Between 2000 and 2009, for instance, the world's population grew from just over 6.1 billion to about 6.9 billion, and it is expected to rise to about 9.3 billion by 2050. Putting this into context, as recently as 1830 the population was only 1 billion (Barry, 2005: 256)

Of course, as Young (1993: 4) points out, environmental problems do manifest themselves at the local level, but 'what appear to be little local difficulties are the visible parts of much more complicated sets of inter-related problems', with regional, national and international aspects. Thus, for many people, the increasing volume of traffic and the building of more roads to meet rising demand causes readily visible congestion and damage to the countryside. It also, though, causes less visible health problems and, even further removed, it contributes to acid rain, increases the level of CO_2 in the atmosphere thereby adding to the threat of global warming and, last but not least, uses up more of the world's precious oil reserves (Young, 1993: 5–6).

Despite the obvious relevance of environmental degradation in an explanation of why environmental protection has become a much more salient idea, we should not over-exaggerate its importance. These objectively defined problems can, of course, be distinguished from a subjective awareness of, and concern for, such problems so that the existence of the former does not necessarily by itself explain why the environment has become an important issue. This is one reason why the social sciences can make an important contribution to the environmental debate, and why it is not merely the preserve of scientists, technocrats or even philosophers – important though their contributions might be (Yearley, 1992: 49, 184–85). This recognition explains why there has been a tendency for environmental science degree courses to be redefined as environmental studies (Young, 1990: 91).

At one extreme, it has even been suggested that the objective conditions are not at all important in explaining the rise of a social problem such as the environment (Kitsuse and Spector, 1981). As Chapter 2 illustrates, some of the standard explanations for the rising popularity of the idea of environmental protection – based on an affluence-induced post-material culture and a post-war occupational shift – do come close to denying the social importance of the increasing severity of environmental problems. While we should obviously not ignore completely the explanatory capacity of objective environmental problems, the opposite extreme – that there is a simple relationship between identifying environmental problems, providing remedies for them and the generation of widespread popular support for their implementation – is equally simplistic.

For one thing, many environmental problems are not directly observable (although their effects might be) or, as in the case of natural resource depletion, not easy to visualise. Such problems are mediated through scientists – whose conclusions are rarely universally accepted within the scientific community – the media and pressure groups. In addition, even though most people have indirect experience of environmental problems and disasters, many of them still remain distant affairs with few immediate effects. For example, even though we were told that the Chernobyl nuclear accident affected us – in terms of an increased incidence of cancer – the effects remain imperceptible and we can comfort ourselves with the somewhat complacent thought (encouraged, rightly or wrongly, by nuclear scientists with a vested interest in the continuation of the industry) that our nuclear safety record is such that a similar accident could not happen here. (The relationship between scientific evaluation of risk and the 'real' world inhabited by the general public is discussed by Beck, 1992.) Likewise, we are told that this or that extreme weather event, or natural disaster, is a product of climate change but we have to take the word of experts for this, and there are other 'experts' who put a different view.

Political thinking and the environment

The debate about the rise of the idea of environmental protection illustrates that there is a social science dimension to the issue, but what is the specifically political character of environmentalism? One, extremely limited, answer is to say that the political dimension consists in the science of policy-making; the means by which the state can, and ought to, act so as to ensure the natural environment is protected. This is certainly a part of what environmental politics is about, but it remains a very restricted definition. To see that this is so, consider a popular definition of politics which is *the process by which groups representing divergent interests and values make collective decisions.* There are two assumptions here. The first is that all societies of any complexity must contain diversity; that humans will always have different interests and values, and therefore there will always be a need for a mechanism whereby these different interests and values are reconciled. The second assumption is that scarcity is also an inevitable characteristic of all societies. Since there is not enough to go around of the goods that people want, there needs to be some mechanism whereby these goods can be distributed.

Seen in this way, how might we characterise a politics of the environment? The best way of beginning to answer that question is to consider under what circumstances a politics of the environment could be said to *not* exist. According to our definition, this is where competing interests and values, as well as scarcity, do not exist. It is for this reason, of course, that Marxists claim that a communist society is one where politics is not necessary. For Marxists, since differences of interests in society centre on the existence of competing social classes, the creation of a classless society offers the prospect of a society based on consensus and co-operation – one in which politics and the state is not necessary. Politics, for Marx, then, is seen in negative terms. It is about class conflict. Political power, as Marx and Engels famously insisted in the *Communist Manifesto* (1976: 105), is 'merely the organised power of one class for oppressing another'. It logically follows from this that, once that conflict is ended through the overthrow of capitalism, there are no competing classes and therefore, by definition, no politics.

Can such a scenario be envisaged in the case of the environment? Well, consider that, in his acceptance speech for the Nobel Peace Prize he was awarded in 2007, Al Gore made the claim that 'the climate crisis is not a political issue, it is a moral and spiritual challenge to all of humanity'. In a similar vein, Tony Blair (2005), in a speech made as British Prime Minister, said that: 'Global warming is too serious for the world any longer to ignore its danger or split into opposing factions on it.' The reason why Gore does not see climate change as a political issue – and that Blair wants to avoid it becoming a political issue – is presumably because they think it is a 'no-brainer'. In other words, they think that climate change will damage everyone's interests because it will destroy the planet. It is therefore in everyone's interests to do something about it, and fast. In other words, there is no political decision to be made.

Clearly, if Gore and Blair are right – that climate change affects everyone equally and will ultimately result in catastrophic effects for us all – then they may have a case that there is not, or should not be, a politics of climate change. In such circumstances, it is in all of our interests to act so as to deal with the problem. It is for similar reasons that during a war, a country's internal politics is put on hold so that the common threat can be dealt with, as in Britain where Parliament voted to suspend elections during the Second World War. However, Gore and Blair are not right about this, because climate change, as with other environmental issues, is as much concerned with competing values, ethics and interests as it is with objective facts.

To see that this is so, note that even if Gore and Blair are right that climate change is a threat to human survival – so that, ultimately, we all stand to lose equally if we fail to act to tackle it – this is not the end of the political character of the debate. I might rationalise that many of the catastrophic effects they – along with many others – think will inevitably follow inaction may not happen in my lifetime, and therefore I might decide to continue exploiting the planet. Others may think that such an attitude is selfish in the extreme. This might be because I am childless whereas others would like to pass on a good quality natural environment to their children and grandchildren. It also might be because others think we have moral obligations to future generations, irrespective of whether we know them personally or not.

In reality, Gore and Blair's assumption – that everyone now living is affected equally by climate change – is also incorrect. The impact of climate change has differed, and will continue to differ, from state to state, and from community to community, and the costs of dealing with it are going to be similarly diverse. It is for this reason, of course, that climate change is a political issue. It is not now really a technical issue. Most, albeit not all, have accepted that the build-up of CO_2 – the main greenhouse gas – in the atmosphere is largely man made, and the solutions to this are well known. Either cut down on the amount of CO_2 emitted, adapt to the consequences of allowing it to continue, or find an effective way of capturing it. So, the reason why these solutions have not, so far, been effectively implemented is almost entirely down to the political character of the issue. To be more precise, it is because the effects of climate change impact upon people, groups, classes, nations and regions very differently. Some countries, regions and localities will be hit harder by the impact of climate change, some groups and classes are more able to deal with the consequences of climate change than others, and some will have to make greater sacrifices to act on climate change than others.

Of course, most environmental issues are not ones in which human survival is at stake, or at least not for the foreseeable future. It is true that, in the 1970s, the environmentalists' case was structured by warnings of imminent catastrophe encouraging, as one stream of thought, a survivalist mentality where the objective imperatives – act now before it is too late – predominated (Ehrlich, 1972; Goldsmith et al., 1972; Hardin, 1968;

Heilbroner, 1974; Meadows et al., 1972). More recently, however, environmentalists have put greater emphasis on the desirability, as opposed to the necessity, of change. Thus, a doom and gloom scenario has been discarded in favour of promoting a society which places a good quality environment above one which worships material consumption. As Robyn Eckersley (1992: 17–21) points out, the Green case is driven above all by an 'emancipatory' ethos, since it is not just telling us that we have to give up our present material standard of living, leaving us to mourn our loss, but it is also telling us how our lives can be enriched by adopting a set of values and institutions which will make us happier and more fulfilled. This appeal to self-interest is coupled with an appeal to our altruistic nature, since a central feature of the radical Green approach is an ethical case for discarding an anthropocentric approach to the natural environment in favour of an ecocentric one which recognises the intrinsic value of nature.

Interests, values and inclusion

To sum up, then, a politics of the environment has to consider competing interests and values. It involves a debate about what kind of society we want to live in. Pollution in general, for instance, does not raise insurmountable difficulties as a technical problem, but it enters the realm of politics precisely because it is not merely a technical problem, but one which causes conflict between competing interests – for instance, motorists, oil companies and road builders versus cyclists, pedestrians and wildlife and, at a supranational level, the developed versus the developing world – which decision-makers must seek to resolve.

In addition, central to a politics of the environment is a debate about who we want to include as morally considerable entities. For this reason, a consideration of ethics is, in my view, central in a book about environmental political thinking. This is in line with the widely held assumption that political and moral philosophy are intimately related (Swift, 2014: 5–6). Political thinking about the environment, to be sure, must focus on central political entities such as the state, democracy and justice (the subject matter of Chapters 6, 7 and 8 respectively). However, the character of a Green theory of politics is, to a large extent, predicated on who is to count morally.

Indeed, a characteristic feature of much Green thought is its challenge to the view that only currently existing humans count. Put simply, a theory of the state, democracy or justice that did not recognise the case for including the interests of future generations and non-humans would look very different from one that did. Environmental political thought (a term that is often regarded as synonymous with Green political thought), then, moves the relationship between the human and non-human world 'to the center of its inquiry' (Gabrielson et al., 2016: 4). As Smith (2016: 105) notes, in more detail, environmental political thought 'attempts to build theories, citizenship, and political rights and duties on a more expansive understanding of the community of justice'. Of course, this assumes that an anthropocentric, or human-centred, ethic is intellectually flawed. Chapters 4 and 5 of this volume seek to test this assumption by asking the question: for whom ought we to protect the environment?

Radical and reformist versions of environmentalism

To make sense of the competing claims or discourses (Dryzek, 2013: 8–13) present in environmental thought, it is useful to identify the constituent parts of reformist and radical perspectives (see Box 1.1). To avoid confusion, it should be pointed out that various terms have been utilised to distinguish these two positions. Dobson (2007) and Porritt (1984) refer to dark and light green approaches (only the former justifying the label 'Green'); Young (1993) and Hayward (1995) prefer to use the terms radical and weak or reformist environmentalism; Naess (1973) coined the terms 'deep ecology' and 'shallow ecology'; Eckersley (1992) distinguishes between ecocentric and anthropocentric approaches, while to add to the confusion, both O'Riordan (1976) and

Box 1.1 Reformist and Radical Approaches to Environmentalism

Reformist	Radical
1. Modified sustainable economic growth/ Ecological Modernisation.	1. Limits to, and undesirability of, economic growth.
2. Large role for technological development as a provider of solutions for environmental problems.	2. A distrust of scientific and technological fixes.

Continued

Reformist	Radical
3. Environmental solutions can co-exist with existing social and political structures.	3. Radical social and political change necessary: either authoritarian (for 'Survivalists') or decentralised and democratic political organisation.
4. Anthropocentrism and a commitment to intra-generational and inter-generational equity.	4. Intrinsic value of nature or, at least, a weaker version of anthropocentrism; a commitment to social justice within human and non-human nature.

Pearce et al. (1993) distinguish between ecocentric and technocentric approaches.

The term 'ecologism' is preferred by some radicals because it signifies the interrelationship between the human species and nature, and implies a non-hierarchical order of things displacing man from his dominant position – both key characteristics of the radical approach. Use of the label ecologism, however, can lead to confusion since the term ecology – first used by the scientist Ernst Haeckel in the 1850s – also describes a branch of biology which studies, in a neutral fashion, the relationship between living organisms and their environment (Heywood, 1992: 247).

The differences between the radical and reformist positions are more easily definable than the terminology would suggest. Each approach contains an economic, political and philosophical perspective. Put simply, the reformist position is anthropocentric or human-centred, holding that protecting the environment is primarily for the benefit of humans. In addition, it suggests that environmental protection can be effectively incorporated within the political and economic structures of modern industrial society, without fundamentally threatening economic growth, material prosperity or liberal democracy.

For the reformists, then, economic growth and environmental protection are not necessarily incompatible objectives. Economic development must be sustainable; it must, in the words of one well-known definition, be 'development that meets the needs of the present without compromising the ability of future generations to meet their own needs' (World Commission on Environment and Development, 1987: 43). One way of defining environmental reformism is to describe it as the politics of catalytic converters, power-station scrubbers and bottle banks. It is therefore an optimistic approach, putting faith in the ability of science and technology to solve environmental problems without fundamentally challenging our institutional and value systems.

Many would regard this as the totality of what is involved in being Green or environmentally aware, but from the radical perspective at the other end of the spectrum much more is required. Indeed, some argue that the reformist approach represents an attempt to depoliticise the environment since exponents try to answer concerns about the environment in a way that makes 'sure that things remain the same, that nothing really changes, that life (or at least our lives) can go on as before' (Cook and Swyngedouw, 2012: 1973). This is, of course, to assume that environmental reformism does leave things as they are, which is disputable. Moreover, it also assumes that attaching a high value to elements of the existing social and economic order – based on economic growth, an anthropocentric ethic and liberal democratic political structures – is somehow not legitimate and also not widespread.

For radical Greens, fundamental economic, social and political change – nothing less, that is, than the creation of a new kind of society with different institutions and values – is required both to deal with the severity of the crisis and to enable humans to live more satisfying and fulfilling lives and to provide nature in general with the respect it deserves. From this perspective, the reformists' tinkering with the structures of modern industrial society – providing a few palliatives to mitigate the worst effects of industrial society – is not enough to forestall environmental catastrophe and represents an inadequate and 'shallow' response to the environmental crisis.

A number of general points can be made about the approaches sketched above. In the first place, it is not being claimed here that this is the only, or even the most adequate, typology available. Barry (1994, 1999), among others, for instance, regards the polarisation endemic in the division between radical and reformist as unhelpful, not least because it tends to belittle, or direct attention away from, the important task of developing a theoretical perspective which can help us to understand the nature of environmental politics in the present, thus enabling us to chart a course which recognises the many obstacles standing in the way of sustainable development.

Dryzek (2013) provides a more complex typology which helps to counter Barry's objection. By distinguishing between 'prosaic' and 'imaginative' discourses Dryzek offers us the possibility of preferring the much more sophisticated 'sustainability' model which, while remaining reformist, confronts directly some of the main radical objections to a more simplistic or 'prosaic' reformism. Dryzek's typology also reminds us that an anthropocentric approach need not necessarily lack a radical edge. As we shall see, anthropocentric theories of environmental justice, for instance, can be very far reaching in their implications. This is the case in terms of their prescriptions for both inter- and intra-generational equity.

The second general point is that the distinction between reformism and radicalism is over-simplistic: an inevitable and necessary feature of all typologies. In particular, it is possible to hold positions in both camps. An acceptance of the Limits to Growth thesis, for instance, does not preclude one from espousing an anthropocentric ethic. Similarly, there is nothing in principle incompatible with holding a belief in radical social and political change while at the same time accepting the role of technological innovation. Third, the typology is incomplete. Even the reformist approach recognises that there is an environmental 'problem' requiring action. Both the radical and reformist approaches are challenged by a so-called 'Promethean' or 'cornucopian' approach which denies the existence of acute environmental problems and has 'unlimited confidence in the ability of humans and their technologies to overcome any problems – including environmental problems' (Dryzek, 2013: 52). This discourse, although the dominant mode of thinking about the human relationship with the natural environment in the last century and for much of this century, was, as Chapter 3 reveals, articulated as never before in response to the challenge it faced from environmentalists in the 1970s. Since then its fortunes have waxed and waned according primarily to the state of the economy (*ibid*: 52-72).

The debate between the reformist and radical strands structures the chapters in this book. Chapter 3 considers ideas relating to political economy. The idea of the limits to growth is critically examined in the context of Promethean denials and the alternative idea of sustainable development. It is argued that of central importance to a political economy of environmentalism is the validity of the claim that limiting economic growth is a desirable, rather than an inevitable, objective. Chapters 4 and 5 deal, predominantly, with the idea of inclusion, since they consider the case for regarding entities other than currently living humans as morally considerable. Following this, Chapters 6, 7 and 8 focus on the political implications of incorporating the interests of non-humans.

Finally, Chapter 9 assesses the environmental credentials of a range of traditional political ideologies before asking both whether the political ideology of ecologism can be regarded as distinct, and whether it ought to be so regarded.

2 THE EMERGENCE OF THE ENVIRONMENT AS A POLITICAL ISSUE

An obvious initial question to ask is why the idea of environmental protection has become so politically salient in recent decades. Intuitively, one might think that the rise of environmentalism as a political issue maps on to the existence of objective environmental problems. However, we cannot just assume that the former accounts for the latter. Indeed, adapting Martell (1994: 4), it is possible to identify three possible types of explanations for the rise of the environment as an important social and political issue. First, there are those that see growing environmental concern as a by-product of cultural and structural factors happening independently of the actual objective state of the environment (reviewed in Lowe and Rudig, 1986: 513–20). Second, there are those that place emphasis on the mediating influences of the environmental lobby, the media and scientists. Finally, there are those that focus on the existence of worsening environmental problems as the key trigger for concern. The chapter concludes, therefore, by documenting the range and effects of these problems. By then, however, it should be appreciated that documenting the objective reality of environmental problems is much more straightforward than explaining why the environment has emerged as an important social and political issue.

Cultural and structural explanations

We might be led by our intuition to assert that objective environmental problems are self-evidently important explanations of rising environmental concern. There are – cultural and structural – explanations, however, which deny this 'common-sense' view of the world and, for this reason, we will start this chapter by examining their claims in some detail.

Inglehart and post-materialism

Inglehart provides the best example of the cultural explanation for the rise of environmental concern, linking the growth of post-material values with affluence (see Box 2.1). In support of the affluence explanation, the development of the environmental movement does indeed seem to be linked with the economic climate. Lowe and Goyder (1983: 16–17), for instance, identify four periods – the 1890s, 1920s, the late 1950s and the early 1970s – when concern for the environment was particularly evident, and these do correspond with the end of periods of extensive economic expansion. This would also seem to explain why concern arose again towards the end of the 1980s as the economy was booming, and dropped off in the early 1990s when recession arrived. As Worcester's research (1997: 162–63) reveals, support for environmentalism is 'correlated negatively with the Economic Optimism Index, which measures people's expectations of future economic prosperity', thereby confirming the view that 'people's concern for the environment rises when they feel economically more secure'. Clearly, then, the economic climate does have an effect on the issues people regard as salient.

Box 2.1 Inglehart and Post-Material Values

Proponents of a cultural change focus on a shift in values which, it is argued, has produced as one of its consequences mounting concern for the environment. Ronald Inglehart (1977; 1990; Inglehart and Rabier, 1986), the best-known proponent of this type of explanation, charts the rise since 1945 of a section of the population expressing so-called post-material values (see Table 2.1). These, he argues, have arisen as a consequence of post-war affluence. As the welfare state and economic success have taken care of many people's material needs in the developed world, they have increasingly turned their attention to meeting non-material goals, one of which is the desire to live and work in a pleasant environment. These values are particularly prevalent in the generations who have grown up since 1945 and so, although post-materialist values are not yet held by a majority of the adult population, there is every possibility that they will become predominant in years to come. The theory, therefore, is not dependent upon individuals changing their values but upon a new generation growing up in an affluent society adopting post-materialist values.

Continued

Table 2.1 Materialist and post-materialist value types in six Western European nations by age group, 1970.

Age group	65+	55-64	45-54	35-44	25-34	15-24
% Materialist	48	45	36	35	31	20a
Post-Materialist	3	7	8	12	14	25a

a:The generation born after 1945.

Source: Inglehart (1990: 49)

On the negative side, there is little evidence that numbers of those holding post-material values have increased as Inglehart predicted they would. Indeed, there is contrary evidence that the Green vote in some countries is getting older, as those earlier post-materialists have not been replaced by equivalent or greater numbers of younger people (Carter, 2007: 98). In addition, one of the pillars of Inglehart's theory can be challenged. Here, it needs to be asked why the satisfaction of material needs leads to post-materialism rather than new material needs. Inglehart derives his ordering of values from Maslow (1954), who argued that needs are pursued in a hierarchical order according to the extent to which they are necessary for survival. Once the basic physiological needs are met, he suggests, other higher-order needs, such as love and esteem, come into play. It is by no means clear, though, why this should be the case. For one thing, what one generation regards as essential will differ from another. At one time, televisions and motor vehicles were regarded as luxuries; now they are common and not to own one through lack of resources is regarded as a form of deprivation. In the future, computer ownership, and access to the internet, will no doubt fall into this category (Martell, 1994: 125).

The new class

The structural interpretation is associated above all with Stephen Cotgrove and Andrew Duff (1980; 1981), who argue that the rise of environmentalism as an important issue reflects the ideological disposition of a new social grouping which has emerged in the post-war period (see also Cotgrove, 1982; Gouldner, 1979). This new fraction of the middle class (NMC) consists of those who work in the non-productive service sector – doctors, social workers, teachers and so on – a sector that has grown extensively since 1945. It is occupation as opposed to affluence, then, which, according to this view, is crucial. Those working in new middle-class occupations are much more likely to espouse post-material values than the private sector working class and middle class because, it is argued, they are insulated from the dominant values of industrial society. Environmentalism, then, is 'an expression of the interests of those whose class position in the non-productive sector locates them at the periphery of the institutions and processes of industrial capitalist societies' (Cotgrove and Duff, 1980: 340–41).

The structural explanation does seem to account for the specific character of the Green social base rather better than Inglehart's affluence explanation. That is, it is not the most affluent sectors of the middle class, but those in public sector occupational groups, who are most likely to be active in Green movements and parties (Doherty, 2002: 226). Persuasive though the structural explanation may sound, however, it has a number of problems. In the first place, many who hold post-material values are not members of the new middle class but are from peripheral segments of society such as homemakers, the retired, students and the unemployed (Norton, 2003: 109). Moreover, there is evidence from the 1980s onwards of a dilution of the NMC dominance of those supporting environmentalism (Norton, 2003: 110; Tranter, 1996).

Second, it is difficult to justify the claim that support for post-materialism among the NMC can be explained by class interests. If this is so, it has to be shown that the NMC stand to gain from policy measures recommended by environmentalists. No doubt this is sometimes true, but not obviously so. Clearly, some environmental campaigns, particularly those that seek to prevent further development in leafy suburbs (the NIMBY syndrome), are motivated by class self-interest (Lowe and Goyder, 1983: 28–30). However, this does not account for the post-materialism of the NMC since those in white collar public sector jobs are far from being the most affluent part of the middle class. Moreover, as Lowe and Rudig (1986: 522) point out, why should those working in the public sector espouse the curtailment of economic growth when it is precisely a thriving economy which ensures the continued existence of the welfare state?

Support for environmentalism among the NMC might, of course, be about values rather than class self-interest. One major objection here, though, is the difficulty of determining causality between occupation and ideals. Even Cotgrove and Duff (1980: 102) admit that 'environmentalists try to choose occupations congruent with their ... post-material values'. If this is the case, the whole theory is devalued since post-material values can no longer be explained in terms of occupation. That is, it is not the occupation itself which causes individuals to hold post-material values, but a tendency for those already holding such values to search for appropriate occupations. This, of course, takes us back to square one and our search for an explanation for the existence of the value shift in the first place! This does not mean that Cotgrove and Duff are prepared to accept the affluence argument of Inglehart after all, since their research still suggests that material conditions are not an accurate predictor of post-material values (*ibid*: 106).

The interdependence of environmental problems

We will come back, in the conclusion to this chapter, to a further analysis of the likely utility of cultural and structural explanations for the rise of environmentalism. In this section, the common-sense explanation – the existence of severe objective environmental problems – is considered. An account of the environmental problems facing the world focuses on the effects of human (or anthropogenic) influences on the natural environment, although it should be noted that natural phenomena – earthquakes, volcanoes, landslides, cyclones, droughts and so on – can have a devastating impact too. To add an additional complication, it is also claimed that severe weather conditions – including droughts and floods – are a product of man-made climate change, and in so far as this is the case, it makes it much more difficult to distinguish between anthropogenic and non-anthropogenic effects. Indeed, a central feature of environmental politics is that the extent and consequences of the human impact on the natural environment are much disputed.

Utilisation of the Earth's natural resources stretches back many centuries, but the crucial turning point was the industrial revolution, which led to fundamental changes in manufacturing activities and resulted in far greater stresses on the natural environment than before. Industrial expansion resulted in rapid investment-led technological innovation, unprecedented population growth and the large-scale movement of people from rural to urban areas and from towns to cities (Young, 1992: 11). Economic expansion has continued apace and received a further boost after 1945, when rising expectations demanded greater material prosperity, and a new financial system was designed to meet these expectations.

Although convenient, it is somewhat artificial and misleading to consider different forms of environmental damage in isolation from others, since the problems are interdependent, a fact illustrated by the

Limits to Growth report discussed in Chapter 3. Thus, one particular process has multi-faceted consequences, and actions to deal with one specific consequence often have knock-on effects causing unintended, and often undesirable, by-products. Both the inputs and the outputs of the industrial process cause environmental damage (Georgescu-Roegen, 1973). Raw materials are transformed into finished products by processes requiring energy. The inputs of raw materials, and the energy sources required to transform them, are usually non-renewable, leading to potential problems of resource depletion. Moreover, the extraction and use of raw materials and energy sources may well cause pollution. The manufacturing process not only causes pollution but also land-use problems. In addition, by creating a large urban population, industrialism also requires an efficient, intensive and environmentally (and ethically) dubious farming sector to provide adequate food supplies.

The end-products of the industrial process have to be transported by road, air or sea, and this has implications for energy consumption, pollution and the fate of the countryside (not to mention the species of flora and fauna residing there), parts of which may have to be destroyed to build the necessary transport infrastructure. Once these products have outlived their usefulness, they have to be disposed of. This again raises issues of land use, water and air pollution, and the fate of the humans and non-human species which inhabit the areas utilised.

Many of the products of the industrial process are damaging to the environment when in everyday use, and none more so than the symbol of the modern environmental crisis, the motor car (see Box 2.2). Without doubt, too, the emergence of climate change as an environmental problem has reframed other environmental issues. The impact of the car culture has traditionally had local and regional environmental impacts – such as land use and pollution. It now has global implications, however, as a major contributor to overall CO_2 emissions. Moreover, climate change has become a key cause of loss of biodiversity, as well as impacting upon food and water availability. Two other examples are pertinent. First, nuclear power has been traditionally opposed by the environmental movement, but it is now back on the political agenda (with, in particular, both the British and the French Governments re-establishing their nuclear programmes) because it is a carbon neutral alternative to the burning of fossil fuels – a point emphasised, controversially, by James Lovelock (2006), the doyen of radical environmentalists. Second, climate change reinforces the need for population growth to be stemmed as the more people there are the greater the greenhouse gas emissions will be.

Box 2.2 The Motor Vehicle and the Environmental Crisis

It is easy to see why the motor vehicle has become a symbol of the environmental crisis. Above all, it helps to illustrate why many environmentalists argue that there are no small-scale, piecemeal solutions to the crisis. Producing motor vehicles uses valuable raw materials and energy; vehicles consume energy and pollute the environment while in use; they require roads to run on; they cause urban congestion, thereby reducing the quality of life for many; and, finally, they must be disposed of when no longer needed.

Seen in this context, the solutions proposed and acted upon seem entirely inappropriate. While the use of lead-free petrol and the fitting of catalytic converters has stopped some of the worst emissions, they do not deal with the other consequences of car use. While cars continue to be produced, they still reduce air quality, consume energy and require roads. What is more, they still emit CO_2, the gas which is regarded as a primary cause of global warming. Diesel cars have become popular in recent years, and governments have encouraged consumers to purchase them, because they produce less carbon dioxide and are more fuel efficient, and therefore cheaper to run. Ironically, though, they produce much higher levels of air pollutants, particularly nitrogen oxide. The extent of the air pollution caused by diesel cars – thought to result in over 30,000 early deaths a year worldwide (*The Guardian*, 15 May 2017) has been underestimated because car manufacturers – most notably Volkswagen – have been found to have falsified emissions testing data (Harvey, 2016).

Governments have begun to act in response to these problems. The British Government, for instance, has announced that both petrol and diesel vehicles will be prohibited by 2040. The move to electric cars, however, raises its own problems (such as the impact on the electricity supply system) and will still have many environmental problems (such as the pollutants caused by tyres and conventional land-use problems). Given this, environmentalists can be forgiven for the claim that the only genuine solution is a drastic reduction of road vehicles rather than more environmentally friendly ones.

Such has been the impact of humans on the environment that geologists have now suggested that a new 'Anthropocene' epoch in planetary history is upon us (Crutzen and Stoermer, 2000). The transition from the 'Holocene' (the 12,000 years of stable climate since the last ice age, during which all human civilisation developed) marks a period where humans no longer adapt to changes in their natural environment. Rather, it is humans who are directly responsible for altering the environment and, in the process, degrading it (Carrington, 2016; Garrard et al., 2014). The aim of this part of this chapter is to sketch out the changes that humans have wrought.

First-generation environmental issues

It is useful to divide an account of environmental issues into first- and second-generation problems. The former category includes air and water pollution, declining biodiversity, resource depletion and, as a cause of these – at least in part – population growth. Each of these constitutes an enormous topic, and it is not the intention of this chapter to provide anything other than a brief sketch (for a good introduction to the issues see Middleton, 2013).

In terms of pollution, harmful chemicals, gases and particulates are regularly emitted into the atmosphere during various stages of the industrial process, ranging from arsenic – emitted through oil and coal combustion and glass manufacture and linked to lung and skin cancer, to manganese – emitted from power stations and during smelting and linked to Parkinson's disease (Simpson, 1990: 56). A high proportion of the total number of harmful substances are emitted from fossil-fuelled power stations and vehicle exhausts. In combination, these are responsible for so-called photochemical smogs and the phenomenon of acid rain (McCormick, 1989). Over two-thirds of the Earth's surface is covered by water and, as it is essential to human life, protecting it against pollutants is crucially important. In industrialised countries, the biggest water pollution problem is not now organic wastes but industrial and agricultural wastes, including radioactive chemicals, nitrates, heavy metals and oil. The main problem areas are: (1) nitrates and phosphates from the use and disposal of fertilisers and detergents; (2) dangerous organic chemicals, most notably polychlorinated biphenyls (PCBs), involved in the manufacture of paints and plastics, and pesticides; (3) heavy metals such as mercury, lead, arsenic and aluminium; and (4) oil.

Oil spills are one of the most visible signs of pollution, but the spectacular accidents widely reported in the media account for a relatively small amount of the total quantity of oil deposited in the sea, the rest coming from routine discharges from ships cleaning out their tanks, natural seepage, industrial effluents and the consequence of war, such as the deliberate sabotaging of Kuwait's refineries by Iraq in the 1991 Gulf War. When the *Torrey Canyon* ran aground in 1967, it deposited 117,000 tonnes of oil off the coast of South West England, but this spill rather pales into insignificance when one contrasts it with industry figures which report that approximately 5.65 million tonnes of oil were lost as a result of tanker incidents from 1970 to 2009 (according to the International Tanker Owners Pollution Federation (ITOPF) website www.itopf.com/information-services/data-and-statistics/statistics/).

The number of plant, non-human animal and insect species with which we share the planet is startling. Indeed, such is the incredible variety of life on Earth and the difficulty of determining the numbers that we simply do not know how many species there are. Excluding plants and vegetation, estimates of the total number of living species vary from between 5 and 30 million. Only about 1.5 million have so far been identified. It is probable, therefore, that some species have disappeared (perhaps recently) without us knowing they ever existed. What we do know is that in the last 2,000 years, humans have exterminated about 3% of the world's mammal species, one-half of these losses occurring since 1900 (Regenstein, 1985). The crucial point to make here is that human activity is responsible for extinction rates that would appear to have accelerated way beyond their 'natural' level (Tuxill, 1999: 97).

Despite the signing of the Convention on Biological Diversity at the UNCED in 1992, there appears to be little improvement in extinction rates. The International Union for the Conservation of Nature's Red List of Threatened Species reported, for instance, that in 2010 no less than 21% of the 5,490 recorded species of mammals, and 12% of the 9,998 recorded species of birds, were threatened, and the number of critically endangered species of mammals and birds had increased from 169 and 168 in 1996 to 188 and 192 respectively by 2010. By 2016, of the 82,954 plant and animal species assessed, 23,928 were classed as threatened in some way. Of the 5,107 species considered critically endangered, around half are plants, with amphibians, molluscs and fish also having large numbers of species on the brink of extinction (www.iucnredlist.org/). The Harvard entomologist Edward O. Wilson estimated, conservatively, that three species are being extinguished every hour (*The Guardian*, 18 February 1995).

There are many threats to species, some of them natural, but many of them deriving from human activity. Hunting, for instance, has seriously eroded the numbers of many species of whales and poaching has done likewise for African elephants and rhinos. The greatest threat, though, is not caused by deliberate direct attacks on non-human species but by the destruction of the habitats on which they depend, particularly for agricultural purposes. Tropical forest depletion is far and away the main cause of species extinction. As a consequence, then, of the uneven distribution of the Earth's biodiversity, the issue (as with environmental protection in general) is inextricably linked with the relationship between North and South and the latter's desire to develop. Finally, just as human beings are affected by climate change, other species will also count the cost in terms of the spoliation of habitats and a reduction in food supplies. The classic example is the plight of the polar bear, whose ability to hunt for food is being hindered by rapidly decreasing sea ice.

Clearly, then, excessive species extinction does occur as a result of human activities. The question to ask is does it matter? The first answer is to say that it matters because nature has inherent value which we ought to respect. We will return to this moral question in Chapters 4 and 5. More pertinent for the present discussion is the anthropocentric dimension – that protecting biodiversity matters because it is beneficial to us. Here we are on safer ground because, in many instances, this is clearly the case (Pearce et al., 1993: 98–100; Abramovitz, 1997: 95–114), although balancing the interests of endangered species with those of indigenous human populations in the developing countries is not easy (see Box 2.3).

Box 2.3 Anthropocentric Reasons for Protecting Biodiversity

There are a number of convincing reasons why it is in the interests of humans to protect biodiversity.

- First, there is the aesthetic pleasure we receive from experiencing the majesty and beauty of non-human nature.
- More instrumentally, many plant and animal species are economically valuable to us, in terms, for instance, of tourism and food. For example, there is an economic disincentive to hunt whales to extinction since, with effective conservation measures, stocks can be made to last longer. Indeed, the International Whaling Commission (IWC) was set up with precisely this intention, although in recent years it has been used for different purposes by those with a principled opposition to whaling (see Chapter 4).

Continued

- Genetic variety is also important from a medical, agricultural and industrial perspective (Tuxill, 1999). Many commonplace medicines derive from natural sources – aspirin, the contraceptive pill and cancer-treating agents, to name but a few – and there is every reason to think that future discoveries may greatly benefit humans and other animals. Biodiversity has made these contributions with scientific investigations into only 1% of plant species and a far smaller proportion of animal species (Myers, 1989: 30–32). Maintaining the diversity of the gene pool is also crucial in guarding against disease which could otherwise eradicate plant species vital as sources of food for humans and other animals.
- Finally, there is the more general point that the workings of nature are so little understood that we simply do not know the full consequences of our actions in destroying species. This brings us back to the interdependence of the natural world, that damage to one part of an ecosystem can have unsuspected knock-on effects – a point that was graphically illustrated by Rachel Carson's (1962) classic work on the effects of pesticides (see also Lear, 1998).

Continuing with the interdependency theme, the emphasis on economic growth, powered by the use of fossil fuels, not only leads to pollution and atmospheric changes, but also raises questions about resource depletion. Statistics on the use and life-span of non-renewable resources are all too revealing. For example, it is estimated that humans use the same amount of fossil fuels in one year that it took nature one million years to create. Utilisation of non-renewable resources, and in particular oil, has taken a heavy toll. There is disagreement on so-called 'peak oil', the point in time when the maximum rate of global oil extraction is reached. Even the optimists within the scientific community, however, argue that the peak will be reached by 2020, after which oil production will begin to decline (Jackson, 2009: 8).

There is then an obvious need to focus attention on the conservation of energy and the development of energy alternatives, particularly given the desire for economic development in the developing world. One answer is nuclear power, which at present constitutes only about 15% of the global energy supply (but over a quarter of Britain's energy sources and as much as three-quarters in France). The British Government has planned to sanction the construction of a whole new generation of nuclear reactors. Environmentalists tend to take a dim view of the nuclear option but it does, at least superficially, seem to answer most of the problems. In particular, as stated above, it provides energy without the problem of CO_2 emissions (for a generally positive account see North, 1995: 80–90). Of course, this is not the whole story. There are significant difficulties involved in the disposal of nuclear waste, high capital costs are involved in setting up and running nuclear power stations and there is the ever-present danger of accidents which, as Chernobyl demonstrated, can easily turn into catastrophes (Porritt, 1984: 173–74).

There are many alternative sources of renewable energy – hydroelectric power, wind power, tidal energy, solar energy, geothermal energy and biomass forms of energy. Each has its negative aspects in terms of cost or effects on the environment and, whatever else might be said, it remains the case that a great deal more research is necessary to make any one of them viable (North, 1995: 119–21). There are other barriers too, such as the great vested interests involved in the production, marketing and retailing of fossil fuels. In addition, the large subsidies given to the nuclear industry, largely for political reasons, have created a significantly uneven playing field which makes it extremely difficult for advocates of renewable energy sources to compete effectively. In what many regard as a 'game-changer', however, technological innovation ensured that in Britain, by 2015, onshore wind energy became cheaper than any other source of generating electricity (*Independent*, 7 October 2015) and by 2017 the cost of offshore wind power reached a new low (*The Guardian*, 11 September 2017).

Global atmospheric change

As if the long-standing 'first generation' problems of pollution and resource depletion, discussed above, are not bad enough, in recent years evidence has arisen that human activities have damaged the environment in much more sinister ways. We refer here to ozone depletion and global warming or climate change. Both are genuinely global problems in the sense that all countries, to varying degrees, are responsible for them and affected by them. A global response to them is, therefore, essential if a viable solution is to be found. Individual countries acting alone, or in concert with a small number of partners, is inadequate. In actual fact, ozone depletion has been the subject of a reasonably effective international treaty, and has shifted out of the limelight in recent years, so the focus in this chapter is on the more pressing issue of climate change.

Global warming was first observed in the 19th century. It refers literally to an increase in the temperature of the Earth and is widely thought to be a product of the changing balance of oxygen and carbon dioxide (CO_2) in the atmosphere – other so-called greenhouse gases making a smaller (about 25%) contribution to global warming are methane, nitrous oxide and tropospheric ozone (see Box 2.4). There is no doubt that there is more certainty than there was a decade ago about the existence of climate change and its causes and consequences. However, there are still sceptics. They adopt a variety of positions. Some deny that global warming is happening, although only a very few mavericks still hold to this. Others accept the reality of global warming but deny that its causes are anthropogenic (Singer and Avery, 2007). Still others recognise the existence of climate change, and accept that most of it is man-made, but suggest that the sacrifices required to combat it are so great, and therefore unlikely to be forthcoming, that we should commit resources to ensuring that the developing world in particular is better equipped to adapt to the effects of a warmer Earth (Beckerman, 1992; Lomborg, 2006, 2007).

Box 2.4 Climate Change

Like the glass in a greenhouse, CO_2 acts as an insulator and the more of it there is, the more difficult it is for the sun's warmth to escape. The process of photosynthesis, whereby plants absorb CO_2 and release oxygen, helps to ensure that a balance is maintained. Human activities, however, have put increasing stresses on this natural mechanism, putting more CO_2 and other greenhouse gases in the atmosphere while at the same time reducing the availability of natural 'sinks' for CO_2 through increasing deforestation. Most important of all is the additional CO_2 released through the burning of fossil fuels. In the absence of human activity, the carbon in fossil fuels would remain unused, out of circulation under the Earth's surface or 'congealed' in trees, but fossil fuels, in the form of coal, oil and natural gas, are the mainstay of industrial society and their use has dramatically increased.

Despite the sceptics, there is a scientific consensus that climate change is man-made and has potentially far-reaching negative consequences unless the amount of CO_2 emitted is reduced substantially.

The scientific consensus is that, by 2050, greenhouse gas emissions need to be reduced by about 40% of what they were in 1990 if the world is to avoid the 2 degree Celsius rise in average temperatures (above pre-industrial levels) that is deemed to be the point at which dangerous and irreversible climatic changes will occur. Driving forward the consensus is the Intergovernmental Panel on Climate Change (IPCC), the epistemic community consisting of the world's climate scientists. Sceptics have suggested that the IPCC is a political body with a vested interest in confirming the validity of climate change. Against this, however, it should be recognised that governments have accepted that climate change represents a real problem, and, given the perception that dealing with it will involve economic sacrifices, one would have thought that they would be delighted to use any doubts as an excuse for inaction. Moreover, before the IPCC publishes its reports they have to be accepted, line by line, by hundreds of experts and governmental representatives. The level of consensus reached, therefore, is astonishing.

The regular IPCC scientific assessments over the past 20 years have become increasingly definite in their assertions. In the IPCC's Fifth Assessment Report, published in 2014, it was concluded that most of the observed increase in global average temperature since the mid-20th century is 'extremely likely' due to the observed increase in anthropogenic greenhouse gas concentrations (IPCC, 2014), and that even if the world begins to moderate greenhouse gas emissions, warming is likely to cross the critical threshold of 2 degrees Celsius by the end of this century (IPCC, 2014). The Paris Agreement on climate change, signed in 2015, committed the participants (including, significantly, countries from the developed *and* the developing world), to acting so as to prevent the 2 degree rise. However, the US administration led by Donald Trump has subsequently announced its intention to pull out of the agreement, and a widespread failure of governments to live up to their promises has been reported. In 2016, the concentration of CO_2 in the atmosphere increased at record speed to hit a level not seen for 3 million years (*The Guardian*, 30 October 2017).

Scientists are no less convinced, in the main, about the effects of global warming. Rising sea levels are regarded as a key consequence caused by an increase in the volume of sea water, as the polar ice caps melt, and the expanding of warmer oceans. Since the early 1970s, ice and snow cover has decreased by about 80% and recent evidence shows an increase in plant growth in the Antarctic. The extent of the sea ice in the Arctic reached a new record low in September 2012. Moreover, in a so-called 'feedback loop', the loss of ice cover reduces the ability of the Earth's surface to reflect heat, thereby intensifying warming by increasing the amount of heat absorbed. The result of rising sea levels will be flooding, initially of low-lying areas. In addition, increases in temperatures will affect vegetation and agriculture and, of course, there is a limit to the amount of heat which living organisms can cope with. IPCC scientists have estimated that a 'no action' scenario would lead, by 2050, to a situation where there will be 'about 22 million additional people at risk from hunger because of climate change (over and above that expected because of population change), about 23 million additional people at risk from coastal flooding, and 110 million additional people living in countries with extreme water stress' (*The Guardian*, 2 December 2015). Added to this is the reputed economic cost that will result from failing to act on climate change (see Chapter 3).

If the finger is to be pointed at greenhouse gas emissions, the solutions are clear-cut. One is to use other sources of energy, including nuclear power (which, whatever else may be said about it, does not involve the use of fossil fuels) and renewable alternatives based on wind, water, waves and so on. Another is to use existing sources of energy more efficiently through better insulation of homes and workplaces and restrictions on car use (and the development of electric alternatives) and aeroplane flights. A technological solution is the development of carbon capture and storage, which, if successful, would go a long way towards solving the problem of climate change (although it would not affect many of the other environmental problems described in this chapter) without necessitating cuts in industrial activity (Kunzig and Broecke, 2008; Lomborg, 2010).

Finally, the conservation of forests, and particularly the tropical rain forests, is also crucial. The preservation of the rain forests illustrates well the interdependence of environmental problems. Not only do the rain forests provide habitats for a huge variety of animal and plant species, as well as protecting against soil erosion and the silting of rivers and lakes, but they also perform a vital function as 'sinks' for CO_2. Far from being conserved, however, the rain forests are being cut down at an alarming rate. The problem is not just that forests are felled for timber, but they are often cleared to create agricultural land. The burning of the felled trees not only releases further CO_2 into the atmosphere, thus intensifying the global warming problem, but can also cause severe pollution, and thereby health problems, as witnessed in the forest fires in Indonesia and other places. Moreover, the cattle that move into the space created by deforestation increase levels of methane – also a greenhouse gas.

Conclusion: The truth is stranger than fiction

The bulk of this chapter has been concerned with sketching out the major environmental problems facing the global community. They are formidable and have increased over the past two decades or so. By themselves, these problems constitute a possible explanation for the rise of social and political concern about the environment, but, as we saw, there are those who would challenge the view that there is an automatic link between the identification and popular dissemination of information documenting these problems and the rise of concern about them.

Both the value- and structural-based theories can be criticised, of course, on the grounds that they completely ignore the possibility that it is the deterioration of the environment, and those agents seeking to draw attention to it, which has played a key role in bringing the issue on to the political agenda. Consider this, though. If there is a direct link between the state of the environment, on the one hand, and rising public concern and political action on the other, then we would expect the latter to occur where, in time and space, environment problems were/are most severe. However, this is clearly not always the case. Victorians, for instance, put up with very poor environmental quality, presumably because they thought that the sacrifices were worth it for the economic benefits produced by the industrial revolution. Likewise, in spatial terms, some of the worst examples of environmental degradation have occurred in parts of the world – the old Soviet bloc to give one example – which have not seen the biggest outpouring of social and political concern.

Nevertheless, to 'effectively divorce … environmental concern from ecological problems' (Lowe and Rudig, 1986: 518) is surely to lose sight of much that is important. To support this claim, Carter (2007: 110) provides examples where the growth in support for Green parties is linked with concern about environmental problems. For example, the Greens in Sweden had their highest levels of support in the late 1980s from areas most severely affected by fall-out from the Chernobyl nuclear disaster. To give another example, rising support for Green parties in the 1989 European Parliament election occurred at a time when the political agenda was dominated by environmental issues such as acid rain and global atmospheric pollution.

At the very least, in response to the structural and cultural explanations, we surely have to take into account the role of science, the environmental movement and the media in making the public more aware of environmental problems and the alternative set of values that some environmentalists hold (Hansen, 1993; Martell, 1994: 125–26; North,

1995: 97–110). Important here too is the impact of the expansion of higher educational opportunities in the post-Second World War period (Doherty, 2006: 59–60; Eckersley, 1989: 218).

Having sought to explain the dimensions of what many regard as our environmental crisis, and why it is that ideas about environmental protection have become much more salient in recent years, the rest of this book seeks to explore these ideas in more detail. This task is begun, in the next chapter, by an analysis of the relationship between environmentalism and political economy.

3 THE POLITICAL ECONOMY OF ENVIRONMENTALISM

The economic realm is central to a study of environmental politics. The choices presented would seem to be stark. It is widely thought that huge benefits derive from ever-increasing economic activity. At the same time, economic growth is said by some to be deleterious to the protection of the natural environment, and, if not constrained, will breach the limits of the planet's finite capacity to support human and non-human life. This chapter explores the relationship between economic activity and environmental protection. It documents the perceived benefits of economic growth before going on to examine the challenges to it. A crucial distinction here is between an environmental critique and a wider social critique of economic growth. In the former, the key question is how far it is possible to decouple economic growth from ecologically damaging practices. In the latter, the key question is how far it is possible to decouple economic growth from the notion of prosperity or well-being.

Economic growth and the idea of limits

For over three centuries, the 'big names' in economics have all been convinced that economic growth is the solution to some of the mainstream problems faced by society, whether it be poverty (Adam Smith), overpopulation (Thomas Malthus), distribution (Karl Marx) or unemployment (John Maynard Keynes) (Kerschner and O'Neil, 2015: 250). At least since the end of the Second World War, economic growth – measured in terms of wealth, income, profits and employment – has been the most widely shared objective of public policy. Thus, for Jackson (2009: 3), a growth in GDP has been 'the single most important policy goal across the world for most of the last century'. Growth is seen as natural and hugely desirable. It has been seen as the source of all good, and its absence is regarded as catastrophic. Democratic elections are largely fought over which politicians and political parties are perceived to offer the most credible programme for economic growth – parties of the left and right being united over its desirability. By contrast, environmentalists, and others, who question the wisdom of economic growth have traditionally been regarded as at worst crazy and at best idealistic.

Governments have pursued economic growth with considerable success. It is estimated that by the end of the 20th century, the global economy was 68 times bigger than it was in 1800 (Jackson, 2009: 6). Challenges to the utility of economic growth have been present ever since the industrial revolution transformed what had been primarily agrarian societies. These challenges have centred on the idea of 'limits'; that ever-increasing economic activity cannot continue indefinitely before it starts to encroach on the Earth's carrying capacity, which is finite. Thus, the Earth 'consists of finite ecosystems with fixed stocks of resources, where human population explosion and economic growth threaten to overshoot the limits of these systems or transgress the boundaries of a "safe operating space"' (Dryzek, 2013: 70). This focus on limits is 'a foundation stone of Green political thinking' (Dobson, 1991: 13). The idea of limits can be separated from a more specific critique of the political economy – that it is not industrialism per se which is the problem, but the capitalist economic system in particular (see for instance Barry, 2016). This ecosocialist critique of capitalism is considered in Chapter 9.

The basis of the challenge to economic growth, then, is that there are natural limits to growth and that unless production and consumption levels and population size are reduced to sustainable levels, economic collapse will result. It is better, then, to manage this reduction, be prepared and act on it, or risk being overwhelmed by a crisis which will have unpredictable consequences. As Edward Goldsmith wrote in his *Blueprint for Survival* (a text originally published as an edition of the *Ecologist Magazine* owned by Goldsmith) economic growth will end 'either against our will, in a succession of famines, epidemics, social crises and wars; or because we want it to ... in a series of thoughtful, humane and measured changes' (Goldsmith et al., 1972: 15).

The best-known version of the empirical limits claim is the now-classic study published in book form as *The Limits to Growth* in 1972 (Meadows et al., 1972 updated in Meadows et al., 1992 and 2005). This has been by far the most popular account of limits to appear, but the idea of limits has a long history. One of the first such versions was the now infamous account by the 18th-century economist Thomas Malthus. In a very illiberal fashion, Malthus (known as the 'dismal parson') argued that food supplies were destined to fail to provide for the exponentially increasing population, and that the state should not intervene to rectify the process whereby increased mortality rates bring the two back into balance (Malthus, 1968/1798). John Stuart Mill (1909), to give another example, recommended a stationary or steady-state economy almost a century before Green activists and writers sought to popularise the idea. In addition, immediately preceding *The Limits to Growth* was the neo-Malthusian tone of Paul Ehrlich's book *The Population Bomb* (Ehrlich, 1972) as well as the work of the founders of ecological economics Nicolas Georgescu-Roegen (1973) and Herman Daly (1974).

The Limits to Growth, however, was extremely popular, selling millions of copies and achieving a level of publicity not achieved by other accounts in the limits school. The research for the book was undertaken by academics at the Massachusetts Institute of Technology (MIT), who were commissioned by a group of industrialists, economists and civil servants led by an Italian management consultant Aurelio Peccei, operating in a group called the Club of Rome (for historical background see Hajer, 1997: 80–84). The researchers at MIT (who were neither population biologists nor economists) used a system dynamics approach and a mathematical model (called World 3) which purported to show the relationship and interdependencies between a number of environmental variables – industrialisation, resource depletion, pollution, food production and population – computing the likely result of changes in individual variables (the results of the research are outlined in Martell, 1994: 24–40 and Dobson, 2007, Chapter 3).

The conclusions of this study – setting the agenda for the survivalist stream of thought evident in the 1970s – were quickly grasped by environmentalists as evidence of the catastrophic fate that awaited us unless changes were made to existing levels of production and consumption (for an overview of the survivalist literature see McCormick, 1992, Chapter 4; Dryzek, 2013, Chapter 2) (see Box 3.1).

Box 3.1 The Limits to Growth

The Limits to Growth researchers undertook seven different computer runs, the first assuming continued growth in all of the variables at present trends and the others assuming solutions to one or more of the problems. Two key conclusions were reached. First, the report argued that, because of the interdependence of the

Continued

phenomena being investigated, there are no viable technological fixes available to solve the environmental consequences of growth. Only a significant reduction in economic activity will suffice. Second, it was also suggested that the environmental problems identified (resource depletion, pollution and so on) are increasing in an exponential, as opposed to a linear, fashion, and as a result the crisis point will creep up on us much more quickly than expected (Meadows et al., 1972: 27–29). Continued economic growth, they concluded, would lead to planetary limits being exceeded some time in the 21st century, with catastrophic consequences.

The 'standard run' resulted in industrial collapse early in the 21st century due to resource depletion, but even when, in the second run, a doubling of resource availability was assumed as a result of technological developments, increased industrial output produced unsustainable pollution levels and eventual resource depletion. Even when solutions to both pollution and resource depletion were fed into the model, collapse was still predicted due to food shortages caused by an increasing population and the appropriation of arable land by industry. Solutions to this problem, the report suggested, are equally futile since they have knock-on effects. Diverting capital to agriculture causes a drop in industrial output and even if a solution to food shortages could be found which did not have this effect, the result would be eventual unsustainable population growth, pollution and resource depletion.

The Limits to Growth report was updated in 2005, but the conclusions remained broadly the same. Indeed, the authors reported that they were 'much more pessimistic about the global future than we were in 1972' (Meadows et al., 2005: xvi).

The limits debate continues (although with less intensity than it did in the 1970s), but it has taken off in new directions (Dobson, 2016) and has been reinvigorated by the emergence of so-called 'degrowth', as both a concept and a movement (Demaria et al., 2013). Deriving from France in the 1970s, degrowth is concerned as much with a social critique of growth as with the claim that growth will have undesirable environmental consequences (see below). In addition, more often than not, limits is now explored in the context of so-called 'planetary boundaries', a concept that became current after 2010 (Rockstrom et al., 2009). This school of thought points to a variety of boundaries – for example, CO_2 levels, ocean acidity, biodiversity loss, chemical pollution – each of which constitutes a life-support system, and each of which has a threshold point which, if exceeded, would portend dire consequences. Three of the limits – climate change, biodiversity and the nitrogen cycle – were judged to have already been exceeded by 2009 (see Box 3.2).

Box 3.2 The Limits to Growth and Planetary Boundaries

There are important differences between the notion of planetary boundaries and the Limits to Growth approach. In the first place, the Limits to Growth authors put far less effort into turning their approach into public policy than those associated with the planetary boundaries approach. The former got considerable publicity but did not engage with policy-makers, perhaps because they recognised that their conclusions were unpalatable (Saunders, 2015: 828). Second, there is no emphasis in the planetary boundaries approach on the depletion of extractive resources (such as oil). Third, there are no predictions, explicitly at least, about the likely course of planetary boundaries if economic activity continues unchecked. Finally, and most significantly, no attempt is made to demonstrate the interdependencies between different boundaries (Dryzek, 2013: 35–36).

Limits assessed

Not surprisingly, given its attack on the central objective of government policy throughout the world, the Limits discourse has had little impact on public policy – with the notable exception of China's attempt to control population through its one-child policy (Dryzek, 2013: 43–47). The fact the governments have largely ignored it, of course, does not by itself demonstrate that the Limits discourse is in error. However, the MIT team's work has been criticised on a number of grounds (for a review see Cole, 1973; Martell, 1994: 33–40). In the first place, it is claimed that the computer model was too simplistic, omitting variables which might have produced very different results, and seriously underestimating the capacity of humans to develop technological solutions to environmental problems and to find additional sources of non-renewable resources. The Limits to Growth researchers were therefore accused of choosing a methodology that was likely to produce the outcomes suggested (Blanchard, 2010).

Further, it has been claimed that the conclusions of the Limits research reflected the class and gender interests and ideological bias of the authors, since the conclusions were based on assumptions influenced by an ideological predisposition towards post-material values, which were in turn the product of the comfortable middle-class lives of the authors (for an ecosocialist critique of capitalism in general see Chapter 9). The focus on population control has also provoked the ire of eco-feminists who point to the control over women's lives that this involves (Diamond, 1994). A related criticism was the total absence of any recognition that economic growth in developing countries may be more desirable than in the rich North, thus giving the impression that the poorest peoples in the world should be asked to pay for the excesses of the rich developed world. Thus, the Limits to Growth approach had 'an over-aggregated model of world trends which makes them insensitive to regional differences and the specific situation of Third World countries' (Martell, 1994: 38).

The most ferocious attack on the Limits to Growth report, and subsequent Green arguments based on it, came from a group of scholars who, in the absence of a more appropriate label, can be described as 'cornucopians', or 'Prometheans' after the character in Greek mythology who stole fire from the gods (Beckerman, 1974, 1995; Lomborg, 2001; Maddox, 1972; Myers and Simon, 1994; Simon, 1981; Simon and Kahn, 1984; for an overview see Dryzek, 2013: 52–72). The Promethean position is one which is widely accepted by economists, and indeed, as Dryzek (2013: 52) points out, was the 'taken-for-granted' approach which only required articulation as a result of challenges to it. Moreover, the Promethean arguments were particularly influential in the United States in the 1980s, and again during the period of the Bush administration between 2001 and 2009 (Dryzek, 2013: 59–67; Rowell, 1996) (see Box 3.3).

Box 3.3 The Promethean Position

The Promethean position is centred on a critique of the Limits to Growth team's failure to understand the role of economic feedback mechanisms. In particular, the growing scarcity of a resource results in its price rising and demand for it falling. It is argued that this, in turn, will provoke a search for new supplies of the resource and/or for the discovery or creation of suitable substitutes. The Limits to

Continued

Growth school, it is argued, significantly underestimated the role that technological innovation (and human ingenuity in general) can play in tackling any problems that may be thrown at us. Moreover, in terms of population growth, both Malthus and the Limits to Growth team were wrong to ignore the fact that higher standards of living through economic growth have the effect of reducing the birth rate (Beckerman, 1974: 236–39).

The Prometheans attempt to belittle the Limits to Growth arguments case by providing indicators which, they argue, illustrate the infinite capacity of the Earth to cope with human exploitation and the ingenuity of humans to solve problems that arise. In other words, for the Prometheans, limits do not exist, and the catastrophe predicted by the Limits to Growth researchers has not happened and was never going to happen. They point, for instance, to the declining price of natural resources which is exactly the opposite of what ought to happen if, as Limits to Growth exponents argue, they have become more scarce. Growing food supplies, improving air and water quality as well as longer life expectancy are also invoked to indicate the increasing quality of life for most people. As a result, 'there is no substance at all in the main fears of the eco-doomsters' (Beckerman, 1974: 241). As later Prometheans have put it: 'The limit of sustainability is not a static ceiling but is formed and expanded by human innovation and technological progress' (Lomborg and Rubin, 2002: 43).

Such has been the volume and intensity of criticisms of Limits to Growth ideas that one commentator has argued that their popularity owes 'more to the rise of the environmental movement in general than to their scientific validity' (Sandbach, 1978: 496). This rise, in turn, can be explained by the unparalleled economic expansion of the time, just as the declining interest in the Limits to Growth can be explained by economic downturn from the late 1970s.

In the face of obvious environmental problems, the Promethean approach might seem just too optimistic, complacent, and even reckless. It is certainly the case that some of the data produced by the Prometheans on global trends has been disputed (Dryzek, 2013: 70–71). At the very least, economic forecasting, on which they base much of their case, is an inexact science. Even if we accept the Promethean analysis of human progress to date, there is no guarantee that we will always generate solutions to problems in the way that we might have done in the past (Dryzek, 2013: 69). It also, perhaps, underestimates the degree to which the improvements in overall physical well-being that Prometheans champion are not the product of unregulated human ingenuity, but are a reflection of the considerable success the environmental movement has had in influencing public policy. In this sense, it is important to note that the Limits to Growth report is as much a warning as a prediction, and that, to some degree, that warning has been heeded. Part of this success is undoubtedly also down to a growing recognition, not least among policy-makers, that, contrary to the Limits to Growth position, economic growth and environmental protection are not incompatible.

Sustainable development and ecological modernisation

An indication of the strides environmentalism has made in the past few decades, perhaps, is that the dismissive Promethean response to the claims of the Limits discourse is no longer the norm. Instead, a more common response is an appreciation of the need to tackle environmental problems, but coupled with an insistence that they can be dealt with without the draconian cuts to economic growth suggested by the Limits to Growth school. The pivotal feature of this school, as we have seen, is the assumption of an automatic trade-off between environmental protection and economic growth, so that an emphasis on one necessitates neglecting the other. The reformist's position has become increasingly centred on a denial of this assumption. Instead, it is argued that the key challenge for environmentalists is to demonstrate that sustainable growth is possible; that an emphasis on environmental protection is not incompatible with economic growth and may even enhance it.

The insistence that the relationship between environmental protection and economic growth is not a zero-sum game provides one popular defining feature of the concept of sustainable development. Indeed, this is the dominant interpretation of sustainable development, although the term is sufficiently vague to have many competing definitions, serving the interests or ideological positions of those who espouse them (Baker, 2006: 30–31; Dryzek, 2013: 145–59). For example, from the Limits to Growth perspective, sustainable development can only be meaningful if development is not synonymous with economic growth (Richardson, 1997: 49). In popular parlance, however, the term has been taken to symbolise the attempt to achieve growth that is sustainable. It therefore seeks to reconcile economic growth with environmental protection, and the interests of present and future generations. This is the popular interpretation associated with the Brundtland Report (World Commission on Environment and Development, 1987), which, famously, defined sustainable development as being about meeting 'the needs of the present without compromising the ability of future generations to meet their own needs' (*ibid*: 8) Such a definition 'involves a rhetoric of reassurance. We *can* have it all … no painful changes are necessary' (Dryzek, 2013: 159).

At the level of the developed nation state, an attempt has been made to put flesh on a reformist version of sustainable development through the articulation of what has come to be known as 'ecological modernization' (Barry, 1999: Dryzek, 2013: 165–83; Hajer, 1997; Weale, 1992: 66–92; 113–18). This concept, developed in the early 1980s (Huber, 1982; Janicke, 1985), requires state intervention to restructure the capitalist economy. It has been suggested that ecological modernisation is most likely to flourish in countries which accept the concept ideologically and which are amenable to state intervention. This has allowed decision-makers in some countries – such as Finland, Japan, Holland, Norway, Sweden and most notably Germany – since the 1980s to take action on environmental problems without fearing an economic or political backlash (Wurzel, 2012; Wurzel and Connelly, 2011: 280–82). Since it was accepted that the economic consequences would not be severe, governments were also able to be proactive and take action even when the causes and consequences of environmental problems were not entirely clear. By contrast, ecological modernisation is more likely to be resisted where neo-liberal economics is dominant and the virtues of the free market are extolled. This has been particularly the case in English-speaking nations. Thus, Britain's environmental record can be at least partly explained by the failure of decision-makers to recognise the force of ecological modernisation, and by a chequered history of state intervention (Hajer, 1997: 104–74).

The ecological modernisation position revolves around a number of distinct, but related, arguments, all of which lead to the suggestion that economic growth can be decoupled from ecological damage. In the first place, it is claimed that economic growth is not synonymous with the increased use of non-renewable resources. What is important, therefore, is not reducing economic growth (measured in terms of GDP) but making sure that growth is sustainable. For David Pearce (1993: 4) this means: 'making sure that substitute resources are made available as non-renewable resources become physically scarce and … ensuring that the environmental impacts of using those resources are kept within the Earth's carrying capacity to assimilate those impacts'. This sustainability is to be achieved in part at least through technological developments. To give one example, although Britain's GDP rose by over 30% between 1980 and the end of the 20th century, the amount of energy used stayed about the same (Pearce, 1993: 22–23).

Not only does environmental protection fail to hinder economic growth according to this view, but it can also be a source for future growth. This, it is argued, will be achieved through the production of low polluting goods – 'clean' cars, CFC-free aerosols and so on – which consumers are increasingly demanding. In addition, pollution control technology itself has the potential to become an important source of economic wealth. As a consequence of greater environmental awareness among consumers and investors, companies are being forced to behave in more environmentally responsible ways.

A further claim of the ecological modernisation position is that the trade-off theory overlooks the fact that, although failure to act on environmental problems now may save costs for present generations, these costs do not disappear but are transferred to future generations. Thus, the real cost of damaging nature, according to the UN's report on the 'Economics of Ecosystems and Biodiversity' published in 2010, is at least ten times greater than the cost of maintaining the ecosystem as it is (*The Guardian*, 22 May 2010). In addition, the trade-off theory too often assumes that whatever reduces the profits of manufacturing concerns (as tighter environmental measures would do, at least initially) will result in an overall reduction in a nation's economic prosperity. This, though, is not necessarily the case since it does not take into account the problem of 'externalities' – a phrase that crops up regularly in the environmental literature (Goodin, 1992: 100–104). In the absence of any environmental regulations, factory owners can take the decision to discharge the waste their factories produce into the air and water, thereby reducing costs. For an individual businessman or woman, this may well be a rational decision. Air and water are public goods, external to the profit and loss account. However, the clean-up costs have not disappeared; they are simply transferred to someone else.

Now, it may well be that those who are affected by this pollution are prepared to put up with it in return for the economic benefits the factory provides. It may be, alternatively, that the community decides that they should clean up the pollution and/or insist on the implementation of more stringent environmental regulations so that the externalities are 'internalised' as a cost borne by the factory owner. One of the regular complaints made by environmentalists, as we shall see below, is that the GDP measure of national prosperity is too narrow and should be replaced by a broader indicator of national well-being, which includes the quality of the environment (Connelly et al., 2012: 235–38; Sagoff, 1988). In the context of ecological modernisation, though, the key point is that even if we limit our indicator of national prosperity to straightforward economic factors, allowing factories to pollute indiscriminately will still produce costs since the factories' pollution will damage the economic position of others too. It may, for instance, damage the tourist industry or the fishing industry and it may compromise the health of the people affected, thus diminishing the quality of the workforce and thereby reducing the competitiveness of the economy overall. One study, reported in the *New Scientist* (28 September 1991), for instance, calculated that in Eastern and Western Europe, acid rain causes £200 to £600 billion worth of damage every year to lakes, forests, crops, buildings and human health. A similar scenario has been suggested for climate change (see Box 3.4).

Box 3.4 Climate Change and Ecological Modernisation

Climate change provides an interesting case study of ecological modernisation. It is common currency that acting on climate change will require economic sacrifices. This is the key reason, of course, why a successful climate change treaty has been elusive. This lends support to the Limits to Growth claim that economic growth and environmental protection are incompatible. The ecological modernisation position offers a number of responses to this claim. In the first place, action on climate change can be a source of new Green technology, such as electric cars, and carbon capture. Second, economic growth can continue, without an increase in CO_2 emissions, if non-renewable energy is used more efficiently or if renewable energy sources are further developed. Third, a failure to act on climate change will have economic costs, much of which will be passed on to future generations. This latter argument was the subject matter of the Stern Report (2009), which argued that it is cheaper for developed nations to suffer the economic consequences of cutting CO_2 emissions than to deal with the consequences of dangerous climate change.

The ecological modernisation position seems to offer a viable theoretical framework which can be utilised both as a device to persuade governments to act and as a means of judging how sustainable their response is. One recent version of it is the so-called 'ecomodernism' position associated with a group of scholars and activists (see Box 3.5).

Box 3.5 Ecomodernism

In 2015, a group of scholars and activists produced an Ecomodernist Manifesto (Asafu-Adjaye, 2015). Its basic claim is to affirm the view that 'humanity must shrink its impacts on the environment to make more room for nature' while rejecting the view that 'human societies must harmonize with nature to avoid economic and ecological collapse' (*ibid*: 6). The manifesto shares elements of both the Promethean and ecological modernisation positions. With the former, it shares an optimism that human ingenuity can be employed to solve environmental problems, and that the doomsday scenario painted by the Limits school can be exaggerated. With the latter, it shares a recognition that the state must take action to ensure that technology is put to work in an environmentally optimum way. The manifesto therefore embraces technological innovations such as agricultural intensification, genetically modified and synthetic foods and nuclear power, but, unlike the Prometheans, it emphasises that such technological process 'is not inevitable', as a 'function of market-driven innovation' but requires a 'sustained commitment' to the 'continuing evolution of social, economic and political institutions' (*ibid*: 28).

The evidence is, however, that the use of ecological modernisation by governments has been patchy at best (Carter, 2007: 233–37; Mol and Sonnenfeld, 2000; Wurzel, 2012). In addition, ecological modernisation is concerned with sustainable development in countries in the rich North. One criticism is that it overlooks the possibility that while corporations may behave in a sustainable fashion in developed countries, where environmental regulations are in place, they may relocate their less sustainable activities to the developing South (Carter, 2007: 231).

Finally, advocates of Limits still point out that environmentally responsible growth, wherever it takes place, is still no substitute for low or zero growth and significantly reduced consumption levels. One, largely unacknowledged, reason why this may be so is the distinction between the relative and absolute decoupling of economic growth and environmental degradation. Relative decoupling 'refers to a decline in the ecological intensity per unit of economic output' (Jackson, 2009: 48). That is, for each unit of economic output, technological innovation – as the advocates of ecological modernisation correctly point out – can reduce the impact on the environment. However, as economic growth increases, then the impacts will continue to rise, albeit at a slower rate. Only an absolute decoupling of economic growth from environmental degradation can ensure that economic growth remains within ecological limits. Jackson, for one, thinks it is unlikely that this absolute decoupling will come about through technological advances alone, and that only a reduction in economic activity will suffice.

At the very least, we can say in response that in the real world such a scenario, of reduced production and consumption levels, particularly in the developing world, is perhaps unrealistic. Whether sustainable growth is possible in the long term is an open question. For some radical environmentalists, ecological modernisation is merely a 'rhetorical ploy that tries to reconcile the irreconcilable (environment and development) only to take the wind out of the sails of "real" environmentalists' (Hajer, 1997: 33).

The degrowth social critique

Even if the case for ecological modernisation is proven, and the ecological consequences of limits do not arise, we might still want to say that economic growth is undesirable. This is the social, or well-being, critique of economic growth. The original Limits to Growth authors were concerned only with the ecological consequences of economic growth. A later group of scholars and activists, broadly known as the 'degrowth movement', have questioned the ecological *and* the social undesirability of economic growth (see, in particular, Jackson, 2009; Latouche, 2009 and Victor, 2008. For an earlier example of a social critique of economic growth see Mishan, 1967). In short, the social critique is arguing that 'limits to growth is not a penance to be borne but an opportunity for healthier, happy and fairer lives for all' (Dobson, 2016: 300).

As Jackson (2009: 4–5) points out, to most economists 'the very idea of prosperity without growth is a complete anathema' and yet 'prosperity isn't self-evidently the same thing as economic growth'. Indeed, prosperity 'goes beyond material pleasures. It transcends material concerns. It resides in the quality of our lives and in the health and happiness of our families' (Jackson, 2009: 16). The general aim of the degrowth movement, therefore, is to demonstrate that prosperity is not primarily, or even mainly, a function of economic well-being. Its ultimate goal is the promotion of a simpler, and more democratic, society – one in which people are happier and more fulfilled – which respects biophysical limits. This is an objective which is common to much radical Green thinking.

It is important not to build a straw man here. The degrowth school are not saying that economic growth has had no benefits. Nor are they saying that, particularly in the poorer parts of the world, there can be no benefit from economic growth now and in the future. In this sense, Beckerman's (1974) comment that 'Those who attack some of the most unpleasant features of modern society seem to have no conception of the poverty, squalor, misery and ignorance that was the lot of the mass of the population in the recent past' misses the point somewhat. Rather, the degrowth school are particularly challenging the desirability of continued economic growth in modern developed parts of the world. Economic growth should, as Barry (2016) points out, be replaced by 'economic security' which includes not just a comfortable standard of living but also social security, education, a lower degree of income inequality and income security.

Degrowth scholars point out that surveys of well-being or quality of life (levels of happiness) do not directly map onto levels of economic prosperity. It is certainly the case that in poorer countries, growth does make a significant difference to levels of happiness. It is also the case that survey evidence suggests that 'income and happiness have been positively associated' (Easterlin, 1974: 99–103). However, some countries, such as Denmark and Sweden, record higher levels of well-being than the United States despite the latter having higher income levels (Jackson, 2009: 40). Moreover, overall levels of well-being have appeared to remain static or have even declined, despite being accompanied by increases in economic growth.

There are a number of possible explanations for the apparent lack of association between rising economic growth and levels of well-being. Firstly, degrowth authors point to the diminishing marginal utility of rising incomes, so that once a certain level of economic prosperity is reached, levels of happiness or well-being plateau and do not rise significantly if prosperity increases further (see Layard, 2005). Thus, in most developed societies, reported life satisfaction has remained virtually unchanged over a considerable period of time despite significant economic growth over this period (Jackson, 2009: 40).

Second, while it can be admitted that economic growth has had enormous benefits – improving health, prolonging life and so forth – it is still the case that poverty, inequality and unemployment remain, with all the consequent health and social problems they cause (Barry, 2016: 313–14). Indeed, a recent well-known study has claimed that unequal societies are disadvantageous to everyone, rich or poor (Wilkinson and Pickett, 2009). This might well explain why satisfaction rates are lower in the United States than in European countries where inequality is less pronounced. The concept of GDP does not engage with income distribution. Indeed, it might be argued that economic growth, by continually providing more for everyone (in the so-called 'trickle down' theory), obviates the need for radical redistributive policies. It serves, therefore, as a legitimising mechanism for inequality. Of course, it remains the case that the poorest in developed countries are still relatively wealthy in comparison to those with a subsistence existence in the poorest parts of the world. However, evidence suggests that levels of satisfaction are related to comparisons with those who are closer to us and not to those who are geographically distant (Jackson, 2009: 52–53).

Third, another explanation for the failure of satisfaction rates to keep pace with economic growth might be the so-called 'social limits' to the enjoyment of the proceeds of economic growth (Hirsch, 1977). The argument here is that economic growth increases demand for positional goods – uncongested roads, well-paid jobs, exotic holidays and so on – whose supply cannot be increased to meet rising demand. Economic growth, therefore, fails to bring satisfaction since there are social limits to its enjoyment. The obvious method of dealing with this problem is to increase the price of positional goods, thereby reducing the benefits for many that might, superficially, have accrued from economic growth.

Fourth, whilst economic growth has certainly had some important benefits, it brings with it some negative consequences too. These would include stress, unrewarding work and a weakening of communal ties – all familiar features of modern economically advanced societies. Fifth, economic growth, it is argued, damages the health of democracy. It is commonly believed that governments since the Second World War have pursued a growth strategy as a direct response to the rising materialistic demands of the electorate. An alternative view is that political elites pursue such a strategy and encourage citizens to want material gains, in order to increase their own political power. Economic growth, according to this analysis, is not only environmentally destructive but is also inimical to the democratic health of a society. If this is correct, liberal democratic principles, involving the tolerance of dissent and the preservation of property rights against the state, are well fitted to restrain economic growth (Lauber, 1978). Linked to this is the argument that the more individuals become passive consumers the less they will be active citizens. Indeed, the dominant empirical and, for some, normative, model of democracy in the post-1945 period has been the so-called 'economic' variety whereby elections become simply a means by which political elites, behaving like producers in a market economy, compete with each other to offer benefits to a largely passive, consumerist electorate (Lively, 1975: 88–108).

The case against economic growth, then, is not just about its ecological consequences but also about its social effects. Ending the perpetual search for economic growth and material prosperity is, for the degrowth school, an intrinsic good because it leads to richer and more fulfilling lives. This has echoes in much radical Green writing. For instance, Schumacher (1973: 26) expressed this view well when he wrote that 'prosperity ... is attainable only by cultivating such drives of human nature as greed and envy, which destroys intelligence, happiness, serenity and thereby the peacefulness of man'. Jackson (2009: 36) lists the non-material components of prosperity, which include the ability to give and receive love, to enjoy the respect of your peers, to contribute useful work, to have a sense of belonging, and to have an ethical dimension to life where we can recognise our obligations to others. Certainly, there is considerable evidence of a 'social recession' in modern societies with increasing levels of loneliness, alcoholism and political apathy (see, for instance, Putnam, 2001). Whether or not these negative features of modern society are caused by materialistic values, it is pretty clear that such values provide no solution to them.

At first glance, to accept a less materially comfortable lifestyle – fewer labour-saving devices, fewer, if any, cars, less travel, and so on – does not seem particularly inviting. Greens would reply, though, that, divested of a desire for material acquisitions, people will find fulfilment in presently neglected aspects of life – in intellectual pursuits, in finding themselves spiritually, in enjoying work for its own sake and not for the monetary rewards, and in developing non-dependent and therefore more satisfying relationships with others.

The difficulty of such a claim, of course, is that we are entering the realm of values, and values are subjective. The pursuit of material possessions is intrinsically satisfying for at least some people, and material commodities can furnish us with identity, experience and a sense of belonging. To give one particularly prevalent contemporary example, access to social media requires technological investment in smartphones and tablets, and the possession of the latest models can be a source of great satisfaction and a sense of belonging. Indeed, most of us could attest to the pleasures that material acquisitions have given us at some point in our lives, and no doubt too, many of those without a materially comfortable standard of living throughout the world would attest equally to their feelings of disadvantage.

Moreover, there is a strong case for arguing that a certain level of material comfort is a prerequisite for the development of an individual's spiritual and intellectual life, since many material acquisitions free people from the everyday tasks of physical survival (Beckerman, 1974: 77). As we saw, one of the key explanations for the emergence of the environment as an important political concern is growing post-war affluence, which has enabled people to direct their attention to quality of life issues. Relatedly, as Beckerman (1974: 18) points out, economic growth does not just facilitate the consumption of material goods but also allows for the pursuance of other activities with a spiritual and aesthetic dimension.

Of course, the benefits of economic growth do not consist just in the attainment of material goods. As governments are fond of saying, a growing economy also provides the tax revenues for the state to finance greater help and support for citizens. One important effect of increasing wealth is that people live longer and therefore spend more years not only as unproductive (in material terms) members of society but also in need of health care and social care. However, the curtailment of economic growth would have serious implications for the state's ability to meet the rising demands of an ageing population. Conversely, it might be that a curtailment of economic growth would stabilise or even reverse life expectancy. Neither consequence is particularly palatable.

Conclusion

It will be apparent that the debate about economic growth revolves around disputed interpretations of the empirical data on the link between economic growth, well-being and ecological consequences. It is also partly, as is environmental politics in general, a subjective judgement about values. What we can say is that there is clearly a huge difference between the Promethean claim that we do not have much to worry about, and the ecological modernisation position that recognises that environmental degradation presents a real problem but one which is resolvable if we take specific action. Adopting a risk averse strategy would seem to be the most appropriate course of action. This is particularly the case if we consider the fact that economic growth has, in any case, stalled in recent years – at least when compared to the 1950s and 1960s – and so, therefore, there would seem to be much less to lose pursuing a strategy which does not put economic growth at the top of the agenda. Ultimately, radical Greens have to convince enough people that there is an attractive alternative to a society obsessed by the pursuit of material acquisitiveness. One additional piece of the Green jigsaw that can be used here is the ethical case for protecting the environment, and we turn to this in the next two chapters.

4 ENVIRONMENTAL ETHICS

This chapter explores the ethical dimension of environmental politics. After setting out a few preliminaries, the chapter outlines the key positions in the debate, namely anthropocentrism, sentient-centrism, biocentrism and ecocentrism. Following this, the case for regarding future generations as morally considerable is explored and the main critique of an ecocentric ethic is set out, followed by responses. Finally, the chapter asks how far an enlightened anthropocentrism is an effective ethical vehicle to achieve the goal of environmental protection.

Preliminaries

An environmental ethic asks: for whom ought we to protect the environment? What moral obligations do we have and to whom? Clearly, there is a distinction here between a legal and a moral obligation. The former is enshrined in the law of a particular political jurisdiction, whereas the latter is one that ought to be due even if the law does not enshrine it. It is this latter moral sphere with which this chapter is concerned. Of course, as was pointed out in the introduction to this book, examining who is to be included in the moral universe is a crucial first step in deciding who is to be regarded as important politically.

If we are considering to whom we owe moral obligations, we need to identify the possible candidates. Obviously, currently living humans is one such entity. Most moral and political philosophy has traditionally regarded this group as the only one to which we owe direct moral obligations. The distinctiveness of environmental ethics lies in the fact that some schools at least want to go beyond this group. Future generations is one such entity, but we might also want to consider the claims of sentient non-humans, as well as non-sentient parts of nature whether they be living (in the case of plants and trees) or non-living (in the case of rivers, mountains and beaches).

It is also important to note here two distinctions that are crucial to understanding environmental ethics. In the first place, there is the distinction between extrinsic and intrinsic value. If it is said that an entity has extrinsic value, then it has instrumental value for others. So, to give a hackneyed example, the chair you are sitting on while reading this book has extrinsic value in the sense that it provides you with the means to sit down. We might have a duty not to destroy your chair but this duty is an indirect one in the sense that the duty is owed to you as the beneficiary of the chair and to the individual or institution that owns it, should you not be the owner. By contrast, if an entity is said to possess intrinsic value then it has non-instrumental value. As a result, moral duties we have to an entity possessing moral standing can be owed directly.

To avoid confusion here, and in the interests of a fuller analysis, it should be noted that intrinsic value is sometimes defined in terms of objective value, as opposed to the non-instrumental value definition employed above (O'Neil, 2003). That is, we can make a distinction between intrinsic and instrumental value, on the one hand, and objective and subjective value on the other. In this latter sense, intrinsic value can be either objective or subjective. In the objective sense, an entity has intrinsic value if it possesses value independently of the valuation of the human valuers.

To see what this means in the context of environmental ethics note that it is possible to distinguish between the fact that a moral statement about non-humans is 'by definition a "human-based-interest" statement' and the claim that 'none of my statements about the world have any independence or informative content outside the fact of my humanity' (Vincent, 1998: 133). In other words, the fact that the valuer of the moral status of non-humans is human does not exclude the possibility that non-humans have intrinsic value; a value, that is, which exists independently of the human valuer.

The alternative to such an objectivist view of value is a subjectivist position which holds that the source of all value lies with those doing the valuing. By definition, adopting such a subjectivist position would mean that, since humans are doing the valuing, if they ceased to exist there would be nothing left of value in the world. However, as O'Neil (2003) also notes, the case for intrinsic value defined as non-instrumental value does not depend upon accepting the meta-ethical claim of objectivism. To insist that it does is to confuse the source and the object of value. As he points out: 'That humans are the source of value is not incompatible with their assigning value to a world in which they do not exist.' In other words, the ethical claim that non-humans have non-instrumental value – that they should not be treated merely as a means to the satisfaction of human wants – is independent from the meta-ethical claim that the source of all value is human beings. The fact that we do not have to establish the viability of an objectivist case for value in order to justify the claim that non-humans have intrinsic value (in the non-instrumental sense) is important.

The second distinction that it is important to make at this stage is between moral standing and moral status. Following Goodpaster (1978), the former I take to mean the existence of any degree of direct moral considerability, and the latter I take to mean the degree of moral worth, so that, as Attfield (2014: 43) clarifies, 'moral standing ... is compatible with different degrees of moral significance'. So, if we say that non-humans have moral standing, this is not the same as saying that non-humans have equal moral value to humans. It is possible, in other words, for an entity to have moral standing but to have less moral status or significance than another entity.

Anthropocentrism and its challengers

Preliminaries out of the way, we can now move on to outline the various positions within the environmental ethics debate (see Box 4.1). Pride of place, because of its long history and ubiquity across many disciplines, is anthropocentrism. An anthropocentric ethic argues that the human species is morally superior to non-human parts of nature. According to the anthropocentric position, the rest of nature (including living but not as far as we know sentient parts of nature such as plants and trees, and inanimate objects such as rocks, stones, rivers and mountains) lacks moral standing (in the sense described earlier) and therefore only has extrinsic value. That is, according to the anthropocentric position, non-human nature (or at least non-sentient parts of it) has value but only to the extent that it is of use to us (Passmore, 1980).

Box 4.1 Positions in the Environmental Ethics Debate

Anthropocentrism – holds that only human beings have moral standing, and therefore intrinsic value. Other entities, such as non-human animals, have extrinsic value; that is, they have value but only if of use for human beings (a representative exponent is Passmore).

Continued

Sentientism - holds that only sentient beings - those who are conscious and can experience pleasure and pain - have moral standing (a representative exponent is Singer).

Biocentrism - holds that all individual living things, sentient or not, have moral standing (a representative exponent is Taylor).

Ecocentrism - holds that nature has moral standing, and that collective entities, such as species or ecosystems, have value, and not individual parts of the whole (representative exponents are Leopold and Callicott). The spiritual/consciousness dimension is the defining characteristic of the deep ecology variant of ecocentrism (representative exponents are Naess and Fox).

The basic problem with anthropocentrism (for those who see environmental protection as a priority at least) is that environmental protection is contingent upon human benefits deriving from it. Sometimes it might be beneficial, at other times it might not be (Kopnina, 2012: 236). Environmental ethics is, of course, not necessarily anthropocentric. Indeed, non-anthropocentric environmental ethics has a long history dating back at least to the 19th century when such American writers as Henry David Thoreau and Ralph Waldo Emerson challenged the moral inferiority of the non-human world (Kopnina, 2012: 238). Later work in the post-Second World War period by those such as Leopold (1949), Naess (1973), Stone (1974), Rolston (1975), Callicott (2013) and Routley (1993) reinvigorated this tradition. As a result, there is now a spectrum of views within environmental ethics, from an exploitative anthropocentrism at one end to so-called ecocentrism at the other.

Critics of anthropocentrism seek to extend the moral universe. The so-called 'ecocentric' position, of which deep ecology is a variant, associated most notably with Leopold (1949), Naess (1973), Callicott (2013) and Fox (1984; 1995), seeks to extend the moral universe the furthest by insisting that the whole of nature has intrinsic value in the sense that 'non-human beings are not simply of value as a means to human ends' (O'Neil, 1993: 10). An ecocentric ethic, then, holds that non-human animals have moral standing, that they can be directly harmed morally, and therefore what we do to them should not merely be judged by the benefits to humans of a particular course of action. Equally, though, moral standing for ecocentrism is extended beyond sentient beings to encompass living, but not sentient, parts of nature as well as inanimate objects such as mountains and deserts. An intermediate position – the so-called life-centred or biocentric position – wants to include all living things within the category of entities that can be accorded moral standing or moral considerability (Taylor, 1986).

It is important to note that the characteristics of the ecocentric position apply not only to the relationship between humans and the rest of the natural world but they also impact on how we should behave towards each other. This is drawn out nicely by Eckersley (1992: 56), who points out that 'a perspective that seeks emancipation writ large is one that necessarily supports social justice in the human community' (the relationship between environmental protection and justice amongst humans is considered in Chapter 8).

Two points about these competing positions in the environmental ethics debate are worth raising at this point. In the first place, an ecocentric ethic does not necessarily posit moral equality between the various entities assigned direct moral worth. All it does is to make the claim that non-human parts of nature have moral standing, or are of direct moral worth. This is not the same as saying that these entities have equal moral worth (or status). So, for example, non-human animals may be granted intrinsic value, in the sense that we recognise that they have important interests which matter to them. However, while these interests have to be taken into account, we might also argue that human interests take precedence (see the discussion of animal welfare in Chapter 5). For most ecocentric thinkers, a position which might be called biotic egalitarianism is clearly inadequate. One way of dealing with this is to invoke the moral relevance of sentience. Warwick Fox (1984: 199), for instance, states that:

> the central intuition of deep ecology does not entail the view that intrinsic value is spread evenly across the membership of the biotic community. Moreover, in situations of genuine value-conflict, justice is better served by not subscribing to the view of ecological egalitarianism. Cows do scream louder than carrots.

The second point is that, while biocentrism advocates an individualistic ethic, ecocentrism adopts a holistic approach, concerned with the protection and prosperity of wholes (such as ecosystems, habitats and species) rather than with the well-being of individual parts of the whole. Individualism has been a characteristic of Western thought since the Enlightenment, and has been a major feature of modernity. It denotes the importance of reason and human superiority and separateness from nature and is encapsulated in the deontological tradition as represented, most notably, by the philosophy of Kant (Pratt et al., 2000: 2, 20, 85–87).

Holism in environmental thought reflects the importance of the science of ecology; the insight that without a thriving natural environment the prospects for individual inhabitants within it are decidedly weakened. The most extreme version of holism is so-called 'holistic value theory' (Attfield, 2014: 8). This position accords moral value to entities such as ecosystems or species and suggests that their protection is more important than the protection of the individuals within them. Leopold (1949: 217) offered the classic statement of holistic value theory when he wrote that 'a thing is right when it tends to preserve the integrity, stability and beauty of the biotic community. It is wrong when it tends otherwise.'

As Attfield (2014: 38) points out, 'no value theory focusing entirely on the collective good rather than that of individuals can even begin to be plausible'. However, it should be noted that holistic value theory is not the only version of holism available to us. An alternative, more plausible, way of justifying holism – and one which is consistent with some ecocentric thought – is to say that individual flourishing is dependent on the collective ecosystem so that the extinction of a species can alter the balance of an ecosystem, thereby damaging the interests of all of its members. This position does not, as Attfield (2014: 9) points out, 'deny that ecosystems have great value' but that 'this value arises … from the way that ecosystems facilitate the lives and the flourishing of the numerous individual creatures that comprise them or depend on them'. Of course, this alternative version of holism rejects the view that ecosystems have intrinsic value (and therefore rejects ecocentrism as a valid ethical position), suggesting instead that an ecosystem has extrinsic (or instrumental) value for those individual entities dependent on it (James, 2015: 65–72).

Environmental ethics in practice

It might be thought that the arguments between different branches of environmental ethics are esoteric enough to be suitable only for the university seminar room. This would be mistaken. Much practical wildlife conservation, for instance, can only be understood in the context of competing ethical perspectives (see Box 4.2). We might want to protect wild animals because we recognise their intrinsic value, which requires that we treat them with respect. We might value a particular species, too, for the role it plays in maintaining the balance of an ecosystem. Accompanying these non-anthropocentric reasons for preserving species, though, are a variety of justifications for protecting wildlife which centre on the human interests served by it. Thus, we might appreciate the aesthetic beauty of at least some species, or we might recognise their economic value – in terms of tourism or as a source of economic profit – or medicinal value.

Box 4.2 Whaling and Environmental Ethics

Whaling is a particularly interesting illustration of the ethical conflicts inherent in wildlife conservation. Indeed, it represents an interesting playground for ethicists to explore (see Garner, 2011). The aim of the whaling convention, the International Convention for the Regulation of Whaling (ICRW) (1946) was to place limits - decided by the intergovernmental body the International Whaling Commission (IWC) - on the number of whales that could be caught each year on the grounds, it was thought, that this would help to preserve whaling stocks so they could continue to be hunted. As a device to conserve whales, the ICRW has been largely unsuccessful. As it is an agreement between sovereign states, whaling nations are not obliged to join and, although most have, this is primarily because the ICRW sought to encourage the widest possible membership.

In reality, whale species have tended to be protected only once they have been hunted to the point of extinction. Thus, of the 14 'great' whales, all but the small minke have been exploited to danger levels. Most symbolically of all, the blue whale - the largest mammal ever to have existed on Earth and equivalent in size to 25 adult elephants - was not protected until 1963, by which time its numbers had dwindled to between 200 and 2,000 (*The Guardian*, 13 July 1990). There is no precise way of determining how many whales in any particular species exist and thus what the quota should be. In consequence, political expediency has played a significant part in the IWC's deliberations. It is in the commercial interests of whaling nations to continue exploiting whale stocks to the point of extinction since they can then use the profits to invest in other projects (Cherfas, 1988: 200–203). On ecological and aesthetic grounds, of course, it is a different matter.

Continued

Such was the failure of this policy that in 1986 a moratorium on whaling was introduced to tackle the declining stocks, although whaling for scientific purposes continues. Ostensibly, the moratorium was instituted because whale stocks had declined to such a degree that the continued viability of whaling was in question. In reality, non-whaling nations such as Britain and the United States, who have come to dominate the ICRW, forced through the moratorium, and have voted to retain it ever since, for other, non-economic, reasons. These governments are reacting to the anti-whaling sentiments of their citizens, who, in turn, have been influenced by the campaigning of the environmental movement. To some extent, undoubtedly, the opposition to whaling is based on the perception that whales have intrinsic value and, therefore, they have interests which ought to be protected, but it is also based on the aesthetic value of the mammals which, of course, is equally anthropocentric as the economic justification for conservation. With growing evidence that some whale stocks are now sufficiently large to justify a resumption in hunting, the conflict – between the original intention of the treaty as a means of managing whaling to increase its life-span, and the animal cruelty and aesthetic justifications for continuing the moratorium – has come to a head (see Chapter 5).

Virtually all international wildlife treaties emphasise the benefits to humans of conservation. Some – such as the ICRW – were set up explicitly to protect economic interests. Other wildlife treaties are less explicitly focused on economic benefits, but they remain anthropocentric nevertheless. For example, the Convention on Wetlands of International Importance Especially as Waterfowl Habitat (better known as the Ramsar treaty) explains that wetlands 'constitute a resource of great economic, cultural, scientific and recreational value', while the Convention on the Conservation of Migratory Species of Wild Animals (the Bonn Convention) emphasises that 'wild animals … must be conserved for the good of mankind' (Lyster, 1985: 75–76, 180).

This anthropocentric emphasis also explains why there is a preoccupation with endangered species in national and international wildlife conservation, since human losses (economic, medicinal, aesthetic) are likely to follow from the disappearance of an extinct species but not from the loss of individual animals where this does not threaten the species as a whole. Thus, the central plank of international wildlife conservation is the CITES treaty (Convention on the International Trade in Endangered Species) which came into force in 1975. The aim of the treaty is to restrict trade in some species and prohibit trade in other (more endangered) species. It is not the purpose of the treaty to explicitly prohibit killing or the infliction of suffering on endangered animals, but the treaty does seek to remove one of the major causes of, and justifications for, the killing of members of endangered species.

Future generations

The vast majority of us would agree that currently living humans have intrinsic value and the moral standing that many think accompanies it. We know from our own experience that we can be harmed directly in a way which matters to us. That is not to say that all humans in all circumstances should be treated in the same way, but it does mean that our interests should be taken into account when our actions are evaluated. More problematic is the question of our obligations to future generations – those who are not yet born. Most of us would probably answer intuitively that we do owe obligations to future generations but some challenging objections remain (the issue is discussed in Attfield, 2014: 101–37; Carter, 2007: 45; Connelly and Smith, 1999: 26–31; Hayward, 1995: 140–42; Martell, 1994: 80–85).

The first objection is that it is claimed that we do not owe future generations any obligations because they cannot reciprocate. That is, because they do not exist, they cannot provide anything for us in return for our recognising their interest in inheriting a clean environment. As a result, future generations cannot have rights and we have no corresponding duties towards them. This objection is a product of contractarian theories of justice (see Chapter 8). Since future generations are not able to make an agreement required by contractarian theories, it is suggested they should not be regarded as beneficiaries of the principles of justice decided. John Rawls, a key exponent of the social contract method, recognises that his theory does not find it easy to incorporate future generations and tries to stretch the theory to accord some weight to their interests. He does this by suggesting that participants in the contractural agreement (what he calls the original position) have a motivation to benefit future generations – not because of any feelings of altruism, but because it serves the contractors' interests to do so. Thus, Rawls

(1971: 128) argues that 'we may think of the parties as heads of families, and therefore as having a desire to further the welfare of their nearest descendants'. Participants in the original position still pursue their own interests but these interests are expanded so as to include the welfare of (at least some) future generations.

The problem for Rawls is that, as he recognises, this motivation to act so as to benefit our nearest descendants is only going to cover those future generations nearest in time to the contractors so it is not a principle that will count for generations way into the future. An obvious response to the reciprocity argument is to reject a contractarian theory of justice. Using an alternative not hampered by a need to derive principles of justice from a hypothetical meeting of contemporaries, we can argue that the beneficiaries of justice do not have to be currently existing moral agents. The case for looking elsewhere for an appropriate theory of justice is embellished by the fact that contractarian theories also find it difficult to incorporate existing humans – such as infants and the severely mentally disabled – who are not moral agents.

The second objection to the claim that we owe moral obligations to future generations is the view that we cannot know what future generations will need, and therefore it is impossible for present generations to meet these needs. A related objection is the view that future generations are not out there waiting to exist. We do not know who they are and how many of them there will be. Moreover, in what has been described as the 'non-identity problem' (Parfit, 1984), the identities of future generations depend upon the choices made by those living now. So, to put it bluntly, if we decide to act, say, so as to deplete natural resources then those living in the future will be different from those if we had decided not to take this action. In short, only actual people, and those who are going to exist whatever humans do now, can count morally.

One response to this objection is to ask: does it matter if we do not know which specific individuals will exist in the future, and that we can alter who exists as a result of our actions now? What we do know is that they will be human beings, and, while it is true that needs and wants will undoubtedly change in the future, we do know enough about the basic necessities of life that will not change. Leaving a polluted planet with diminished resources is quite clearly not in the interests of humans in the future, whoever they may turn out to be.

The third objection is the view that, while we should not completely disregard the interests of future generations, the interests of currently living humans should take precedence (Beckerman and Pasek, 2001). We can therefore 'discount' the interests of future generations so that present generations are always regarded as more important, and those who are to come in the future are progressively less important. This would seem to meet the intuition that we should pay more attention to the interests of our children and grandchildren's generation than to those to be born in, say, 200 years' time. The debate about discounting has been most apparent in the ethics of climate change, and is considered in more detail in Chapter 8.

If we do confirm that future generations have a moral claim on us, environmental political theorists then have to consider how those interests are to be represented. Such a task, as Chapter 7 will show, requires us to rethink the nature of representative democracy, and to consider what machinery – legislative inclusion and constitutional protection to name but two – best achieves the goal of politically incorporating the interests of future generations. Some countries – such as Israel and Hungary – have already experimented with the idea of an ombudsman or commissioner whose role it is to represent future generations in the political system (Beckman and Page, 2008: 529).

Evaluating ecocentric ethics

The question of future generations is, of course, still an anthropocentric one. Particularly acute difficulties arise when we seek to justify extending moral standing beyond the human species. The moral status of animals (or non-human sentient beings) is a topic that can be left until the next chapter. Here, we will focus on evaluating the ecocentric and biocentric positions. An evaluation can be conducted at the political level (see Box 4.3) or at the level of ethics. From the latter perspective, one constant criticism of ecocentric ethics is that it is misanthropic (Hay, 2002: 64–65). Much of the force of this criticism stems from the mistaken assumption that ecocentric philosophers are equating non-anthropocentrism with moral equality which, as we saw above, they are not. An ecocentric ethic does not necessarily posit moral equality between the various entities assigned direct moral worth. Nevertheless, the implication of the holistic view associated with Leopold and Callicott is that if humans get in the way of the optimum function of what Leopold refers to as the 'biotic community' then they would seem to be expendable (James, 2015: 54–56).

Box 4.3 Politics and Non-Anthropocentric Ethics

One critique of ecocentrism from the perspective of environmental political thought holds that, while not wrong in itself, ecocentrism does not constitute a political approach. This is because it focuses on what the individual should do and think and fails to direct 'attention toward the material conditions that produce environmental and social crises' (Ray and Parson, 2016: 582). Ecocentric environmental ethics is flawed, therefore, because it is 'abstracted from a social context' and lacks 'a theory of political power' (Hay, 2002: 27–28) which seeks to explain how exploitative power relations result in environmental degradation. From an ecofeminist and ecosocialist perspective, for example, gender and class are the source of environmental problems (and not anthropocentrism per se) and these ideologies offer a political programme to solve them (see Chapter 9).

Another political dimension that ought to be mentioned relates to the question of achievability; a question that has recently been preoccupying political philosophers (Farrelly, 2007). Here, it might be argued that, whatever the ethical merits of a biocentric or ecocentric position, the challenges involved in getting social and political acceptance of them might render an enlightened anthropocentrism a much more realistic proposition, particularly if it can be shown that a non-anthropocentric ethic is 'not necessary to the achievement of the sorts of policy outcomes that flow from deep ecological principles' (Hay, 2002: 59).

Another crucial, and much commented on, difficulty of a non-anthropocentric ethic relates to the moral significance of sentience, or the capacity to experience pain and pleasure. While sentience can be, and has been, utilised to extend moral considerability or moral standing to non-human animals, how can we accord moral standing to non-sentient nature when it is obvious that rocks, trees or mountains cannot be harmed in the same way as sentient animals? Raymond Frey (1983: 154–55) puts the problem succinctly by contrasting the terms 'being harmed' and 'being wronged'. We can quite sensibly talk about non-sentient nature being harmed, just as I can talk about harming my television set by kicking it! But it does not seem sensible to talk about wronging non-sentient nature. Thus, polluting a river is to harm it but since the river only has extrinsic value for those sentient beings who benefit from it, it is only they who can be wronged by polluting the river. At the risk of labouring the point, Peter Singer (1983: 123) expresses the same doubts when he states: 'There is a genuine difficulty in understanding how chopping down a tree can matter to the tree if the tree can feel nothing' (see also Wissenburg, 1993). According to this interpretation, then, sentience is the benchmark for moral standing, and a life-centred or ecocentric ethic is invalid.

There have been some ingenious attempts to overcome the sentience objection to according moral standing to non-human, non-sentient entities. The first point to make is that it is perhaps easier to envisage assigning moral standing to living parts of nature, whether sentient or not, than it is to accord it to the whole of nature, inanimate as well as living. The argument of biocentrics, such as Taylor (1986), is that living things can have interests even though they are not aware of them, so intrinsic value is possessed by those entities who strive for self-maintenance, maturity and reproduction (Hay, 2002: 52). For some environmental thinkers, living things are characterised by the property of 'autopoiesis', which refers to self-production or self-renewal – characteristics which living things possess but which can also be applied to ecosystems or species (Eckersley, 1992: 60–61).

It might be argued that, intuitively, we do recognise the intrinsic value of living things. To illustrate this, a version of the 'last man' thought experiment can be employed. The classic example of this within environmental ethics was provided by Richard Sylvan (then known as Richard Routley) in 1973 (Routley, 1993; see also Peterson and Sandin, 2013). Put simply, consider a nuclear war which wiped everyone out except one (dying) man and a healthy tree. Would it be right for the dying man to chop down this tree? A negative answer would reveal some acceptance of a value for the tree independent of its use for humans, or, for that matter, other sentient beings. That is to say, if you answer that the dying human is not morally entitled to cut down the tree (and, in my experience, many students do not think it is justified) then you are adopting a non-sentient-based non-anthropocentric ethic.

One problem with the biocentric ethic is arguably its impracticality. If all individual living things have moral standing we would have a huge number of moral judgements to make and we would have to consider the prospect that we are 'doing wrong pretty much all the time' (James, 2015: 46). Another problem with the justification for the life-based ethic, however, is that it is not clear how the emphasis on interests sets living things apart from inanimate objects (Hayward, 1995: 67). It is, after all, still in the interests of a television set not to be kicked even if it is not aware of it.

Ecocentric thinkers, of course, do not want to stop at living things. However, assigning moral standing to the whole of nature, and not just living parts of it, would seem to be more difficult still. One well-known approach central to the deep ecology position is the development of an ecological consciousness which requires a change from a position based on a 'code of conduct' to one based on a 'state of being' (Dobson, 2007: 28–42).

Authors in this tradition, such as Fox (1995), Rodman (1983) and Matthews (1991) reject the rationalistic moral extensionist project, in which an attempt is made to demonstrate that nature has intrinsic value, in favour of a psychological approach which documents the need for a wider view of the self (a 'transpersonal ecology' in Fox's words, an 'ecological sensibility' in Rodman's and an 'ecological self' in Matthews'). That is, only if we are able to see that we are not separate from nature but part of it, and enriched by it, will we identify with it and appreciate the need to protect it. Protecting nature, therefore, becomes protecting ourselves. This is related to a virtue ethics approach to environmental ethics, which points to the link between human flourishing and the flourishing of the non-human world (O'Neill, 1993).

It has not been lost on some environmental thinkers (see Carter, 2007: 25; Hayward,1995: 71) that the arguments of Fox and others may, ironically, not take us beyond anthropocentrism and, indeed, can be taken to be a defence of anthropocentrism. This is because these arguments seem to amount to saying that if I am part of the environment, it is in my interests to try and protect it. If so, this would seem to be little different from a moderate anthropocentrism, and therefore open to the objection (explored below) that protecting the environment remains contingent on it being in human interests to do so, which, of course, it might not.

Moreover, as Dobson (2007: 40–42) points out, little attention has been directed to the issue of how this new consciousness is to be generated. Deep ecologists deny that reasoned argument can fulfil this task and say that only 'rhetorical strategies that reach beyond reason to passion' (Dryzek, 2005: 196) will ultimately suffice. Asking humans to 'think like a mountain' or 'howl like a wolf', however, is not likely to take us very far. But without an effective means of showing how and why people should relate to nature, and see themselves as part of it, the force of the non-anthropocentric environmental ethic is minimised.

An enlightened anthropocentrism?

One might reject an ecocentric or biocentric ethic on the grounds that – by relegating humans from the centre of the moral universe – it is not intellectually credible. Even if we accept the moral desirability of an ecocentric ethic, however, the difficulty of establishing such a position academically, coupled with the inevitable problems of persuading decision-makers and the general public of its validity, makes it worthwhile asking whether the achievement of Green objectives necessitates, politically and strategically, such a case being made in the first place (see Box 4.3).

It is clearly the case that anthropocentrism need not necessarily be synonymous with the exploitation and degradation of nature. Eckersley (1992: 33–47), although an advocate of an ecocentric ethic, for instance, does recognise that there is an anthropocentric continuum on which different strands can be placed according to the weight they place on the exploitation of nature (see also Fox, 1995: 160). Thus, the resource conservation type emphasises conserving natural resources to enable their continued exploitation, whereas human welfare ecology focuses on the human benefits deriving from a clean and unpolluted environment and preservationism stresses the aesthetic value of preserving wilderness areas from development.

While Eckersley still regards ecocentrism as preferable to even the more environmentally enlightened anthropocentric approaches, an increasing number of scholars have argued that such moral extensionism is not an essential part of a radical Green position. Goodin (1992: 26–41), for example, seeks to argue that the value of nature consists in its very naturalness and that we derive satisfaction from knowing there are forces of nature operating independently of us and with a history of their own. Goodin recognises that this remains anthropocentric because the value of nature still derives from the satisfaction we get from it and not from something internal to it. Nevertheless, he insists that there can be a gradation of views from anthropocentrism to ecocentrism and that his position is towards the ecocentric end. His green theory of value, therefore, entails the proposition that nature must have value 'in relation to us' but that this is different from saying that nature only has value 'for us'.

Dobson (2007: 42–46), Vincent (1993: 254–55), Barry, (1999, Chapters 2 and 3), Hayward, (1995 and 1998) and Norton (1991) all concur with Goodin's espousal of an enlightened anthropocentrism. Dobson distinguishes between weak and strong versions of anthropocentrism while Vincent develops a continuum bounded by a 'light anthropocentrism' and a 'hard-nosed exploitative anthropocentrism'. Dobson's weak or light version is 'human centred' because only humans can place value on things but this is different, he argues, from the illegitimate strong, exploitative or 'human instrumental' version which involves seeing the non-human world purely as a means to human ends. Much the same reasoning is behind Hayward's distinction between 'human chauvinism' and enlightened self-interest. Finally, Barry (1999: 8, 27) seeks to show that 'a self-reflexive ... as opposed to an "arrogant" ... anthropocentrism, can secure many of the policy objectives of ecocentrism' without the need to enter into deep ecology's '"fundamentalist" complexion which is a hindrance to convincing non-believers to support its political aims'. For these authors, then, an enlightened anthropocentrism provides the necessary protection for nature required by the radical Green position.

The problem with these defences of anthropocentrism is that they still involve nature's value being seen in terms of human benefits. While it is quite correct to say that these benefits can be conceived in terms of the protection of the natural world this is a subjective account, and as such is vulnerable to competing human claims which may be exploitative. As Michael Saward points out 'although Tasmanian mountains may be inspiring and uplifting, the New York skyline may be no less so' (Saward, 1993: 510). So, if Goodin tells me I should not cut down trees because he derives satisfaction from knowing that they are there, I may respond by simply saying that I disagree! That is to say, without an argument to the effect that I should not cut down trees because to do so is to wrong them, then his Green theory of value would appear to have no particular advantage over mine, particularly if I can point to the advantages that will accrue to humans as a result of cutting the trees down.

Of course, there *are* strong human prudential grounds for protecting the environment. We saw in Chapter 2 how environmental degradation causes enormous problems for humans and may even put the survival of the species in the balance. However, it may be possible to drive a considerable distance down the road of environmental degradation before human survival is put at stake. Radical Greens want the brakes put on long before this point is reached, but it is unclear that an anthropocentric approach, however enlightened, is the best means of achieving this ambitious objective. Warwick Fox (1995: 186) nicely sums up the problem of relying on an anthropocentric ethic when he writes that:

> The upshot of this is that even if one wins the battle in regard to preserving this area or that species, one is contributing to

> losing the ecological war by reinforcing the cultural perception that what is valuable in the non-human world is what is useful to humans. If the next battle in which one is engaged concerns an area or a species that is relatively 'useless' to humans then this battle becomes that much harder to win.

Against these arguments challenging the validity of enlightened versions of anthropocentrism, however, it is also the case that the acceptance of an ecocentric ethic does not guarantee that, in the event of a clash, the interests of nature will take precedence over those of humans. This is because an ecocentric ethic only establishes that non-human nature has moral standing; that it is worthy of moral consideration. In other words, it does not tell us that we should protect non-human nature, only that we should consider its interests before deciding on a course of action. There is a crucial distinction, therefore, between moral standing and moral equality or between an entity being morally considerable and the degree of moral significance it possesses (Elliot, 1995: 13).

Conclusion

In this chapter, we have outlined and critically examined the various positions within the environmental ethics debate. An ecocentric ethic, of course, is associated with the radical Green position. Not surprisingly, this ethic has had relatively little impact on the way in which the environmental debate has been couched in the mainstream political arena, although, as subsequent chapters will show, environmental political theorists have taken on board the political consequences of an ethic that removes humans from the moral pedestal. However, the challenge it provides has hastened the attempt to develop an enlightened anthropocentrism. There are compelling reasons to suggest that the human-prudential grounds for protecting the environment are strong and varied enough for us to afford to jettison, or at least not be too preoccupied about, the need to show that the whole of nature does have moral standing. Those ethicists who adopt a non-anthropocentric (life-centred or ecocentric) ethic argue that it is indispensable, since establishing the inherent or intrinsic value of entities other than humans gives them a prima facie expectation of respectful treatment, as well as political inclusion.

5 ANIMAL ETHICS

This chapter explores the ideas current in the field of animal ethics. The first part outlines the positions and approaches in the animal ethics debate. The evaluative contours of the debate between these respective positions are then provided, with particular focus on the animal rights vs welfare and animal rights vs utilitarianism debates. Thirdly, the chapter explores the more recent literature in animal ethics that seeks to differentiate between the moral obligations we owe to wild and domesticated animals. Finally, the often conflictual relationship between animal ethics and environmental ethics is outlined, and the possibility of a reconciliation considered.

Why animal ethics?

The inclusion of a substantive chapter on animal ethics in a book on Green or environmental thinking requires some explanation. As we shall see, a number of prominent environmental ethicists have regarded their concerns as very different from, and at odds with, the stock-in-trade of animal ethicists. This is partly because a good deal of the animal ethics literature relates specifically, or principally, to domesticated animals: those on farms and in laboratories and in our homes. Conventionally, such animals are regarded by environmental ethicists as 'man-made' constructs and, therefore, not properly part of the 'natural' environment. In addition, animal ethics tends to adopt an individualistic and rights- or welfare-based framework, where moral standing is attached on the basis of sentience, or some other combination of cognitive capacity. By contrast, environmental ethics, as we saw in the previous chapter, tends to be holistic, is focused on the non-human world in general, and not just non-human animals (hereafter just animals), and, in some guises at least, does not recognise sentience as the primary benchmark of moral standing.

Three points can be made in response to the claim that animal ethics has no place in a book about Green politics. Firstly, even if we exclude domesticated animals as a proper object of study for the student of environmental ideas, the bulk of animals are, in any case, wild-living (Horta, 2010), and very much part of the natural environment. Secondly, while domesticated animals might not, in themselves, be regarded as part of the natural environment, their existence, and the way in which humans choose to treat them, has a considerable impact on the natural environment. For example, aspects of modern industrialised farming have had, and continue to have, an extremely deleterious effect on the environment. Finally, it is, perhaps, too conservative a step to exclude animal ethics from an account of environmental political thinking. After all, central to a study of environmental politics is a claim that the interests of *all* living non-human entities, domesticated or otherwise, ought to be considered, even if, in some versions, those of humans are regarded as more important. Moreover, it is also the case that the individualistic focus of animal ethics may well provide important insights into how we ought to regard the status of wild animals.

The contemporary debate in animal ethics

Up until the 1970s, with a small number of exceptions, the moral worth of non-human animals was widely regarded, in theory and practice, as inferior to that of humans. This inferior moral status was challenged from the 1970s by philosophers invoking rights-based, utilitarian and contractarian frameworks. This was accompanied by a reinvigorated and radicalised animal protection movement emphasising grass roots campaigning and an animal rights ideology (Garner, 2004).

Prior to the 19th century, animals were commonly regarded, in theory and practice, as having no moral standing. This was typified, philosophically, by Descartes' assertion that animals lack sentience and Kant's claim that animals are not moral agents in the sense that they are not capable of recognising right from wrong, or capable of participating in moral agreements (Descartes, 2012; Kant, 1965). For those who hold that animals have no moral standing, the only duties we have towards them are indirect ones. That is, harming animals only becomes morally significant if it harms the interests of other humans, such as in the common argument that those who are cruel to animals are likely to be inclined to treat humans in the same way. This alleged link, it has been suggested, was the rationale for much of the early 19th-century animal protection law (Ritvo, 1987).

For the vast majority of contemporary moral philosophers (or at least those who have written about the matter), animals are now regarded as sentient, possessing some direct moral worth, and are therefore owed something from us. Of crucial importance in transforming the intellectual climate was the advent of utilitarianism in the 19th century. Jeremy Bentham (1948) set the scene for this transformation in the moral status of animals when he wrote, in an oft-quoted passage, that 'the question' of moral status is not 'Can they reason? Nor, Can they talk? But, Can they suffer?' The utilitarian position, that sentience provides the benchmark for moral standing, has had a considerable influence on debates about the moral worth of animals leading to the acceptability, in theory and practice, of the dominant animal welfare ethic.

The central feature of the dominant animal welfare ethic – what might be called the moral orthodoxy – is an insistence that humans are morally superior to animals, but that, since animals have some moral worth, we are not entitled to inflict suffering on them if the human benefit thereby resulting is not necessary. The principle of *unnecessary suffering*, therefore, can be invoked if the level of suffering inflicted on an animal outweighs the benefits likely to be gained by humans. The political philosopher Robert Nozick (1974: 35–42) provides a concise but admirably effective definition of animal welfare when he writes that it constitutes 'utilitarianism for animals, Kantianism for people'. Sacrificing the interests of animals for the aggregative welfare, then, is permissible providing that the benefit is significant enough, but treating humans in the same way is prohibited whatever the benefits that might accrue from so doing.

Challenges to the moral orthodoxy

This animal welfare orthodoxy began to be challenged from the 1970s. Indeed, this era marked an astonishing flourishing of academic interest in animal ethics. Much of the post-1970 academic literature sought to challenge the moral orthodoxy. The academic debate was, initially at least, dominated by two philosophers and two ethical approaches, the Australian Peter Singer and the American Tom Regan, and utilitarianism and rights theory respectively. There is also a third strand, which I will not dwell on in this chapter, which is drawn from the social contract tradition in Western political thought. This has centred on the work of John Rawls (1971), a central figure in contemporary political philosophy. Rawls reinvigorated the social contract tradition, associated most notably with Locke and Hobbes, to develop principles of justice. He explicitly excludes animals as beneficiaries of a theory of justice because, lacking moral agency, they are not able to participate in a contractarial agreement. Some philosophers, most notably Mark Rowlands (1998, 2009), have sought to adapt Rawls's social contract in a way that allows for the inclusion of animals, therefore providing a third 'strand' of the challenge to the moral orthodoxy (for a discussion of the contractarian approach to animal ethics see Garner, 2012).

These three strands of the challenge to the moral orthodoxy can be regarded as the first wave of contemporary animal ethics. Since then, the animal ethics debate has fragmented in a variety of ways. For example, new approaches – utilising the virtue ethics tradition, on the one hand, and the notion of capabilities, on the other – have been applied to animals (Hursthouse, 2000; Nussbaum, 2006). In addition, a so-called 'political turn' in animal ethics has emerged (see Box 5.1), along with a new field of Critical Animal Studies (see Taylor and Twine, 2014).

Box 5.1 The Political Turn in Animal Ethics

The field of animal ethics has become increasingly populated with political, as opposed to moral, philosophers (see Cochrane et al., 2016; Garner and O'Sullivan, 2016). This raises the question: is it possible to distinguish a political, as opposed to moral, account of animal ethics? For Donaldson and Kymlicka (2011), the political turn refers to the claim they advance that animals ought to be incorporated in the polis as citizens (see below and Chapter 8). While Donaldson and Kymlicka do offer an illuminating new account of animal ethics, it would be wrong to claim that it represents the totality, as opposed to one part, of the political turn. That is, to claim that all theories that deny animals this or any other political status or membership in the polis do not qualify as 'political' theories, would seem to be mistaken.

According to the analysis provided by Cochrane et al. (2016), only the justice strand of the ethical dimension represents both a unifying strand of the political turn literature, and enables us to distinguish this literature from that properly described as being concerned with ethics. All political turn texts, it is argued, 'seek to explore how political institutions, structures and processes might be transformed to better serve the interests of animals' (Cochrane et al., 2016: 10). The stress on justice can be contrasted with the traditional animal ethics emphasis on the moral realm. Whereas the latter focuses on the individual, the former's raison d'être is the state or broader political authority whose role is to enforce the principles of justice.

Animal ethics has also taken a relational turn. Traditionally, animal ethics has used what Palmer (2010: 5) calls a 'capacity-oriented' approach. According to this approach, moral status or worth is granted on the basis of the possession of some capacity or other, whether it is mere sentience or greater cognitive capacities. A major alternative approach is the adoption of a relational ethic (see below). In a pure sense at least, the relational ethic holds that the moral status of an individual is based not upon her capacities or interests but upon the relationship she has with others. The relational approach has its roots in the feminist care ethic tradition (Gilligan, 1983). Care ethicists have argued that we should value emotion and sentiment over reason, and that our moral obligations should derive from the relationships we develop and not as a result of impartial rules. Applied to animals, then, our duties to them are based, not on the grounds of impartial and abstract principles, but on our duty to care about those animals we forge relationships with (Donovan and Adams, 2007).

The overriding problem with relational ethics in general, and the care ethic in particular, is that if we adopt a theory which grounds moral duties, in this case of care, in the relationships we forge, it is difficult to see how moral duties can be applied to those with whom we do not forge a relationship. In the case of animals, for instance, some animals are likely to fare better – companion animals rather than farm and laboratory animals for instance – and it could be argued that the protection of animals is *less* likely to be ensured by an ethic which focuses on partial relationships alone, since our relationships with members of our own species are likely to be stronger than the ones we develop with non-humans. It would seem intuitively mistaken, because morally arbitrary, to deny moral entitlements to animals (and indeed humans) with whom we do not have a relationship, irrespective of their cognitive capacities. Having said that, one area where a relational ethic might be useful (in combination with a capacity-oriented ethic) is in distinguishing our moral obligations to wild and domesticated animals respectively (see below).

The Singer and Regan challenge to the moral orthodoxy

What is significant about the first wave of animal ethics is that it uses previously existing and conventional traditions of thought. The first, in book form at least, was the work of Peter Singer (1990). Singer is an exponent of utilitarianism, the heir of Bentham. Singer suggests that we ought to treat like interests alike (the equal consideration of interests), and since at least some animals are sentient it is morally illegitimate to prioritise human interests in not suffering on all occasions. In some situations, indeed, there may be a case for saying that animals can suffer more than humans and therefore their interests should prevail. This, Singer implies, is the conclusion Bentham ought to have arrived at, but failed to for whatever reason.

Singer's utilitarian position was challenged not only by critics of the attempt to justify an enhanced moral status for animals, but also by those philosophers who argued that Singer's position did not provide a firm enough foundation for the enhanced moral status he sought to establish. Central to this strand of thought was Tom Regan's animal rights position (Regan, 1984). Regan seeks to show that at least some animals are what he calls 'subjects-of-a-life', possessing enough mental complexity to be morally considerable. Because at least some animals have these capabilities, Regan argues that they have a welfare which is capable of being harmed not only by inflictions of pain and deprivation, but also by death since it forecloses all possibilities of finding satisfaction in life. The key point is that rights are not granted to animals by virtue of their sentience (although this was in fact Regan's early position) but because they are what Regan calls 'subjects-of-a-life'; beings with considerable cognitive capabilities. Regan's position is therefore abolitionist. Because animals have a right to life and liberty, there is a moral prohibition on their use by humans *irrespective of the ways in which they are treated*.

Not surprisingly, perhaps, debates *between* the Singer and Regan camps tended to overshadow critical responses from exponents of the moral orthodoxy. Two dimensions of this debate are particularly important. In the first place, Singer bases the moral worth of animals on their sentience and is prepared to accept that (some) humans have cognitive capacities (such as self-consciousness) which may mean that they are of greater moral value (having, for instance, a greater interest in continued life) than non-humans. Regan, on the other hand, bases the moral worth of animals on a range of capacities they have which make them what he calls 'a subject of a life'. Regan is therefore much more prepared than Singer to accept that animals have a moral worth broadly equivalent to that of humans.

The second area of debate surrounding Singer's position is the aggregative character of utilitarianism. Singer is an act utilitarian. Act utilitarianism is a theory of ethics which states that a person's act is morally right if and only if it produces at least as much happiness or pleasure, or the satisfaction of preferences – for the actor and others – as any other act that the person could perform at that time. As such, when judging the ethical permissibility of a particular use of animals he is required to weigh up the total number of preferences, including preferences to go on living, which would be satisfied if that use is prohibited. The problem for Singer, as noted by his critics (see, for example, Frey, 1983; Regan, 1980), is that if we factor in all of the consequences likely to follow a particular improvement to the way that animals are treated the improvement may not prove to be justified on utilitarian grounds. Thus, Singer's claim that such a utilitarian calculation does result in farming animals for food or using them as experimental subjects being deemed objectionable morally is dubious, to say the least. Assuming that we can realistically hope to weigh up all of the preferences involved, there would seem to be numerous human preferences that would be served by continuing these practices. In the case of animal agriculture, for instance, it is clearly not the case that all we have to do is simply weigh up the most fundamental interests of animals against the most trivial interests of humans. Rather, to what might be regarded as a trivial interest – the fact that many humans like the taste of meat – must be added less trivial human interests such as the undoubtedly significant economic costs of the elimination of factory farming or of all forms of animal agriculture.

Defending the animal welfare ethic

Prior to the work of philosophers such as Singer and Regan, the moral superiority of humans had largely been taken for granted. Now, it had to be defended. Traditional defences – based, for example, on claims about human possession of an immortal soul – seem unlikely to succeed in a more secular age. Defenders of human moral superiority are dependent upon being able to show convincingly that humans have morally significant characteristics not possessed by non-human animals. As Steinbock (1978: 247) states: 'We do not subject animals to different moral treatment simply because they have fur and feathers, but because they are in fact different from human beings in ways that could be morally relevant.' The gist of this defence is that humans, it is typically said, are rational, self-conscious, autonomous persons, able to communicate in a sophisticated way and to act as moral agents. This is often converted into a shorthand claim that humans are persons while animals are not (the personhood position). From this empirical evidence, it is argued, can be derived a justification for regarding humans as morally superior to non-humans, and therefore providing a defence of the moral orthodoxy (see, for example, Cohen, 1986; McCloskey, 1987; Steinbock, 1978).

Animal ethicists who dispute the moral force of the personhood position have responded in a number of ways to the claim that humans, because of their status as persons, should be regarded as morally superior to animals. Firstly, it has been suggested that at least some animals do possess elements of personhood. As we have seen, this is the gist of the position adopted by Regan who argues that at least some animals are what he calls 'subjects-of-a-life', possessing enough mental complexity to be morally considerable. These capabilities are:

> Beliefs and desires; perception, memory, and a sense of the future, including their own future; an emotional life together with feelings of pleasure and pain; preference – and welfare – interests; the ability to initiate action in pursuit of their desires and goals; a psychophysical identity over time; and an individual welfare in the sense that their experiential life fares well or ill for them, logically independently of their utility for others (Regan, 1984: 243).

The problem with this position, however, is that most philosophers would argue, with some justification, that even the most intellectually able animals do not have the characteristics of personhood to the same degree as the average normal adult human. That is, although Regan is surely correct to claim that at least some animals do possess more than mere sentience, it is still difficult to establish that the cognitive characteristics possessed by animals are anything remotely approaching those possessed by most humans (see Frey, 1987). As a result, animals will always lose out in a direct application of the personhood argument, if the personhood argument is applied without any further adaptation.

It is correct to say that some non-human species clearly have a greater claim to personhood than others, non-human primates being the prime candidates. This is the rationale behind the Great Ape Project, a movement which has sought to campaign for an extension of rights beyond the human species (Cavalieri and Singer, 1983). However, important though it may be to acknowledge that more needs to be done to protect the interests of the 'higher' non-human animals, all this recognition does is to create a new moral boundary, above which are most humans and those non-human species who possess at least some of the characteristics of personhood, and below which are the vast majority of other non-human animals. It does not, in other words, provide a justification for the kind of broad application of rights to animals that is usually regarded as the aim of animal rights advocacy.

The second response to the personhood position is the assertion that not all humans are persons. Thus, in the so-called 'argument from marginal cases', it is asked if we persist in justifying the exploitation of animals on the grounds that they are cognitively less able than humans, then what should we do with those humans – the seriously mentally disabled in particular – who do not themselves have the characteristics of personhood? Consistency would seem to demand that we either exploit human mental defectives as well as animals, or that we exploit neither animals nor marginal humans (see Dombrowski, 1997) (see Box 5.2).

Box 5.2 The Argument From Marginal Cases

The argument from marginal cases has produced a burgeoning literature (see Garner, 2013, Chapter 9). Those opposed to it have responded in a number of ways. Firstly, it has been argued that infants can be excluded as marginal humans on the grounds that they are potential adults. There is, however, a question mark against the moral importance of potential, one implication being that it makes abortion, and even the use of contraception, a serious wrong (McMahan, 2002: 192). In addition, it has limited application, only applying to infants and not to other marginal humans.

Secondly, it has been argued that the argument from marginal cases underestimates the capacities of marginal humans; that those of even a severely damaged human are still greater than the most mentally developed non-human (Cohen, 1986: 15). That is, most, if not all, marginal humans still have cognitive characteristics closer in kind to 'normal' humans than they do to most species of non-humans. What can be said here is that the psychological capacities of some animals compare favourably with those of some marginal humans. For example, anencephalic infants are born without the capacity for consciousness and those with advanced Alzheimer's disease have little capacity for thought and perception to the point that they have ceased to exist as a person (McMahan, 2002: 44). What is also true, however, is that – although we do not think that

Continued

it is justifiable to experiment on or eat such marginal humans - it still remains the case that we would not think it appropriate to grant them exactly the same moral status as non-marginal humans.

Finally, philosophers have adopted a relational argument which seeks to challenge the use of cognitive capacities by proponents of the argument from marginal cases. For example, McMahan (2002: 217-20) suggests that it is the special relations we have with marginal humans - primarily as relatives - from which we can derive moral reasons to treat them more favourably than animals with similar psychological characteristics. The problem with this argument is that marginal humans are dependent for the existence of the full panoply of rights on relations they have with other, non-marginal, humans. In other words, those marginal humans - such as orphans or those with callous and unfeeling carers - who do not have such relationships do not, according to the relational position, possess the same moral status as non-marginal humans (Gunnarsson, 2008: 309). The arbitrary nature of this allocation of moral value would seem problematic.

The third response to the personhood argument is the – in my mind at least – stronger argument that the mere fact that animals are sentient ought to result in animals being accorded a higher moral status than the animal welfare ethic allows. This suggestion is made by those who adopt an interest-based theory of animal rights (see, most notably, Cochrane, 2012a). As we have seen, the abolitionist assumption derives from Regan's claim that to use animals, irrespective of how they are treated while being used, is to infringe their right to respect which, in turn, derives from what Regan sees as their inherent value. Such an account depends upon the employment of a particular, Kantian, approach to rights which holds that the function of rights is to establish arenas within which individuals can exercise choices (see Hart, 1967; Steiner, 1998: 233–301).

There is, however, an alternative theory of rights, based on interests. The interest theory of rights holds that the function of rights is to uphold individual well-being (Raz, 1986). The possession of an interest therefore leads to a duty on others to ensure that the right, the existence of which follows directly from the existence of an interest, is upheld. Advocates of an interest-based theory of rights are not suggesting that all interests can be translated into rights, only that rights derive from interests.

It may be that we decide that some, relatively trivial, interests are not strong enough to be translated into rights whereas others clearly are, or that some interests ought not to be promoted because they harm others (Cochrane, 2012a: 42–43).

Adopting an interest-based theory of animal rights, however, means that Cochrane can reject the idea that animals are persons but, at the same time, this does not mean that he has to reject the idea that animals do not possess any rights. In particular, while animals may not have a right to liberty, they do have an interest in avoiding suffering (and therefore a right not to have suffering inflicted on them) and so in cases where the use of animals does infringe this right it ought to be prohibited. This interest-based theory allows for the adoption of a more nuanced theory of animal rights. In particular, it enables us to morally separate personhood from suffering. If humans and animals are capable of suffering in an equivalent manner then this capacity of animals should not be downgraded on the grounds that humans also possess the characteristics of personhood. The effect of this analysis, then, is to 'decouple animal rights from animal liberation'. The consequence of accepting what Garner (2013) has called the 'sentience position' is that many of the current practices justified by the animal welfare ethic become morally unacceptable. For instance, inflicting suffering on animals in order to provide food would, in almost all cases, no longer be sanctioned, because it is not necessary to sustain human life. Moreover, those laboratory procedures on animals that were not designed to preserve human life would also be morally prohibited.

Animal ethics and wild animals

Animal ethicists have increasingly turned their attention specifically to the moral status of wild-living animals. The starting point here is the conventional animal rights position which holds that negative rights – rights of non-interference – apply to both wild and domesticated animals. However, as Palmer (2010) – most notably – pointed out, traditional approaches in animal ethics find it difficult to account for the so-called 'laissez faire' intuition whereby it is permissible morally to leave wild-living animals to their fate while it is unacceptable to neglect domesticated animals in the same way. In other words, the according of negative rights seems more appropriate in the case of wild-living animals.

One way of reconciling animal ethics with the laissez faire intuition, Palmer and others suggest, is to utilise relational ethics. The problem for a capacity-oriented ethic is that it cannot justify the according of differential rights (positive rights for domesticated animals and negative rights for wild animals) on the grounds of psychological capacities alone. By contrast, using a relational ethic can justify leaving wild animals alone, and therefore avoid the need to get embroiled in debates about whether we ought to intervene to prevent the suffering of wild animals not least when this suffering is caused by other animals, as in predation. By contrast, domesticated animals, because of the relationship we have with them, can expect us to intervene in a positive fashion to prevent and alleviate suffering, at least when it is not against our interests to do so (see Box 5.3).

Box 5.3 The Relational Ethic in the Work of Donaldson and Kymlicka

Donaldson and Kymlicka (2011), too, adopt a version of relational ethics. In an innovative analysis, they argue that it is useful to map our obligations to animals through the utilisation of citizenship theory. In other words, the moral worth of animals is cashed out, at least in part, through their membership of political communities and the relationships they have with us. Donaldson and Kymlicka envisage three categories of animals, informed by a relational ethic based on citizenship theory. Domesticated animals, those that are part of our societies, are equivalent to co-citizens, and have certain particular rights because of their relational status with humans. Those animals that live among us as but who are not domesticated, so-called liminal animals, are equivalent to co-residents who do not have the rights of full citizenship. Finally, genuinely wild animals are equivalent to separate sovereign communities which ought to be regulated by norms of international justice. In short, only those domesticated animals who are citizens have certain particular positive rights because of their relational status with humans. By contrast, genuinely wild-living animals are equivalent to separate sovereign communities and the assumption here is that they should be left alone by other, human, communities, and maybe given habitat rights as suggested by Cooke (2017) and Hadley (2015). Only the adoption of a relational ethic allows us to make this distinction, since only a relational ethic enables us to differentiate between wild-living animals (where positive duties of assistance are thought not to be required) and domesticated animals (where they are thought to be required).

In response to the use of relational ethics, it might be argued that the intuition – that we should not intervene in the wild – is mistaken. We might not want to go as far as engaging in the kind of human engineering of the wild that Nussbaum (2006) seems to think is necessary when she makes the claim that the natural world must be replaced by the just world. However, given the appalling level of suffering that wild-living animals endure, it does look odd for animal ethicists to regard it as ethically unproblematic merely on the grounds that most of it is not directly caused by humans (Horta, 2010). Moreover, as Cochrane (2012b) points out, we do already recognise a right to protection from predation in the case of some wild-living animals. Preventing cats from killing birds, for example, is a common practice, and not regarded as counter-intuitive in the slightest. In many cases, we may think that intervening to protect wild animals from suffering is too expensive, ineffective or damaging ecologically. The key point, though, is that we have not just assumed in these cases that wild animals should be left to their own devices. Rather, we have concluded after much deliberation that it is the right course of action.

Animal and environmental ethics: the differences

Environmental ethicists, for the most part, adopt a very different approach to animals than the one followed in the animal ethics literature. More often than not they focus on wild-living animals, and typically, as pointed out in Chapter 4, environmental ethicists also adopt a holistic approach, As we saw in the previous chapter too, environmental ethics (just like animal ethics) is not a united discipline. A spectrum of positions can be identified, ranging from anthropocentrism at one end to ecocentrism at the other. It is difficult to reconcile any version with animal ethics, although for different reasons.

From an anthropocentric perspective, it is only morally justified to protect animals when a significant human benefit is served by it. Such an anthropocentric view implies that animals have no moral standing. That is, they have no intrinsic moral worth. This is consistent with the position on animals taken in the past by philosophers such as Kant, but it is at odds not only with animal rights or liberation, but also with the animal welfare ethic, the latter regarding animals as having moral standing if not as much moral significance as humans.

In practice, as we also saw in Chapter 4, much wildlife conservation is driven by anthropocentric, or human-centred, motives, and there are some notable real-world clashes between anthropocentric environmental ethics and sentient-centric animal ethics, not least in the case of whaling. Of course, environmental ethics is not necessarily anthropocentric. The fact that non-anthropocentric environmental ethics accords moral standing beyond the human species makes it consistent with all post-Kantian animal ethics. Both traditions, then, hold that non-human animals have moral standing, that they can be directly harmed morally, and therefore what we do to them should not merely be judged by the benefits to humans of a particular course of action. However, even non-anthropocentric environmental ethics is significantly at odds with animal ethics. At a general level, these differences stem from the fact that much of the animal rights literature accepts the characteristics of the Western ethical tradition (predominantly liberal and individualistic) but argues that non-humans have been arbitrarily, and therefore illegitimately, excluded. On the other hand, the non-anthropocentric environmental ethics tradition began from the assumption that what was needed was a new ethic, not beholden to any previously dominant stream of thought (Jamieson, 1998).

Two contrasts between animal and environmental ethics are particularly relevant. In the first place, moral standing for ecocentric thinkers is extended beyond sentient beings to encompass living, but not sentient, parts of nature as well as inanimate objects such as mountains and deserts. Such a project, of course, has to dispense with mental complexity and sentience as the benchmark for moral standing. Secondly, most non-anthropocentric positions in environmental ethics share with anthropocentrism a holistic focus, emphasising the moral importance of whole ecosystems or species or the concept of diversity itself, rather than the individuals within ecosystems or species.

It is the holistic focus that has been a particular source of difference, and even conflict, between the two sub-disciplines. The conflict is particularly acute in the case of 'holistic value theory' where moral value is accorded to entities such as ecosystems or species rather than to the individuals within them. Holistic value theory is clearly at odds with the individualistic rights or liberation approach to animal ethics. As Regan (1984: 361–62) famously pointed out: 'It is difficult to see how the notion of the rights of the individual could find a home within a view that, emotive connotations to one side, might be fairly dubbed "environmental fascism."'

There is also a tendency in ecocentric thought not only to accept the moral acceptability of sacrificing individual wild animals, but also to have a contemptuous attitude towards domesticated animals, which are regarded as human artefacts having a negative impact on ecological diversity. The most extreme example of this is provided by Callicott who, following in the tradition of John Muir – the 19th-century American environmentalist who described sheep as 'hooved locusts' – wrote that:

> Domestic animals are creations of man. They are living artefacts, but artefacts nevertheless, and they constitute yet another mode of extension of the works of man into the ecosystem. From the perspective of the land ethic a herd of cattle, sheep, or pigs is as much or more a ruinous blight on the landscape as a fleet of four-wheel-drive off-road vehicles (Callicott, 1980: 330).

As we saw in Chapter 4, holistic value theory is not the only version of holism available to us. The alternative version, which simply points out that without a viable ecosystem individuals cannot thrive, would seem to seem to be reconcilable with the individualistic focus of animal ethics. However, even though this version of holism does recognise the moral importance of individuals, it is still at odds with the moral status ascribed to animals in post-Kantian animal ethics. This is because the sacrifice of individual interests (including, presumably, those of humans), is justified according to non-anthropocentric holism if the result is that the bulk of the entities in any one ecosystem or species survives. In practice this has led, most notably, to support among ecocentrics for hunting, a practice which animal rights advocates, and many animal ethicists who are not rights advocates, regard as morally unacceptable. Similarly, this version of holism does not regard vegetarianism as compulsory, as an animal rights position does; indeed, meat eating may be compulsory from the perspective of ecocentrism if it can be shown to be beneficial to the overall well-being of the members of a species or an ecosystem. Indeed, as Callicott (1995: 55) rightly points out: 'Meat eating may be more *ecologically* responsible than a wholly vegetable diet.'

The contrast between environmental holism and the individualism central to animal ethics is most stark when considering what they prescribe in the case of endangered species. An advocate of the former is likely to argue that ecological integrity demands that the protection of members of an endangered species is more important than the protection of a larger number of animals from a non-endangered species. For an animal rights advocate, on the other hand, the fact that some animals are members of an endangered species is irrelevant, and the choice must be to save the more populous group since by so doing more individual rights are being protected. As Regan (1984: 359) points out: 'The rights view is a view about the moral rights of individuals. Species are not individuals, and the rights view does not recognise the moral rights of species to anything, including survival.' As a result, Regan continues, 'that an individual animal is among the last remaining members of a species confers no further right on that animal'.

Case for a reconciliation?

Although the differences between animal and environmental ethics are significant, it should also be noted that there are sources of unity too. It would be wrong, in the first place, to exaggerate the differences. For example, the strand of environmentalism – often described, variously, as ecologism, deep ecology, or dark green ecology – does, like the radical positions in animal ethics, remove humans from the moral pedestal. There is a common enemy here in the form of a hard-nosed anthropocentrism which regards all parts of the natural world as exploitable for our benefit. Those who dump chemicals in waterways, and are responsible for deforestation and the extinction of species such as whales and elephants, are treated with contempt by those in both (animal and environmental ethics) camps. In this sense, notions of love, compassion and a desire to end oppression and exploitation unite both non-anthropocentric environmental and animal ethics (Donovan and Adams, 2007; Milton, 2002; Nibert, 2002).

Secondly, differentiating between our moral obligations to domesticated and wild-living animals is a way of reconciling the two positions. We saw earlier in the chapter that some animal ethicists have sought to do this by invoking a relational ethic (Donaldson and Kymlicka, 2011). Significantly, a similar move has also been suggested by an environmental ethicist (J. Baird Callicott) who was, as we also saw, originally very hostile to individualistic animal ethics. His later, more conciliatory, position is based on his recognition that we can have moral obligations to individual domesticated animals as a result of their membership of a mixed human/animal community. The moral worth of domesticated animals, in other words, derives from the relationship they have with humans. Wild animals are 'not members of the mixed community', however, and as 'members of the biotic community are subject to the duties and obligations of a biotic community or "land ethic"'. As members of this holistic biotic community, he continues, the most fundamental 'fact of life' is 'eating … *and being eaten*' (Callicott, 1988: 164).

The third possibility for reconciling animal and environmental ethics is the adoption of a reformulated, more enlightened, anthropocentrism. As we saw in the last chapter, this has been suggested by a number of Green political theorists as a way of combating the philosophical and political difficulty of establishing a non-anthropocentric ethic Such a case is based primarily on an empirical claim that the protection of the natural world – including animals – *is* in the interests of humans. Here, there is certainly some evidence, as we noted earlier in this chapter, of a convergence between human and animal interests. Animals clearly suffer, like humans do, as a result of environmental degradation. Moreover, there are environmental consequences of treating animals badly. As Jamieson (1998: 46) rightly points out, 'one can go quite far towards protecting the environment solely on the basis of concern for animals'. For example, the damage caused to the environment by factory farming is well established (Mason and Singer, 1990: 72–127). More controversially, raising animals for meat under any system might be considered environmentally detrimental. Jamieson (1998: 46), commenting on the production and consumption of beef, sets out the full human costs. Not only is it a 'moral atrocity', he argues, but it also:

> causes health problems for consumers, reduces grain supplies for the poor, precipitates social divisions in developing countries, contributes to climate change, leads to the conversion of forests to pasture lands, is a causal factor in overgrazing, and is implicated in the destruction of native plants and animals.

As a result, Jamieson continues, 'if there is one issue on which animal liberationists should speak with a single voice it is on this issue'.

However, the relationship between human and animal interests is a contingent one. There are undoubtedly cases where protecting animals may serve no particular human interests. Even more significant, there are cases where there is a conflict between upholding the interests of animals and those of humans. Economic development, for instance (which may well have significant human benefits) often does clash with the interests of wild-living animals). In addition, to give another example, environmentalists are very keen to ensure that chemicals are toxicity-tested extensively before they are allowed to be used. But of course this testing is undertaken largely on animals. Only a rejection of anthropocentrism, in all of its guises, would ensure that the interests of animals are properly considered. This would enable the defence of non-human animal species, *even when* there is a human cost of so doing.

Conclusion

This chapter has tried to illustrate that an analysis of animal ethics has a rightful place in a book about environmental political thought. This was justified on the grounds that animals share the natural world with us, that wild-living animals (which occupy a central place in ecosystems) are considerable in number, that animal ethicists have increasingly turned their attention to working out what our moral obligations to wild-living animals are, that the treatment of domesticated animals often has a considerable impact on the environment and, finally, that central to a study of environmental politics is the claim that the interests of *all* living non-human entities, domesticated or otherwise, ought to be considered.

The chapter has revealed that there are important differences between the animal ethics and environmental ethics literatures, and that these differences have been played out in the practical politics of wildlife conservation. This does not mean, however, that the animal ethics approach should be discarded. It has played an important role in challenging the holistic and anthropocentric agenda that has dominated wildlife conservation. Moreover, the individual focus of animal ethics should remind us that species are made up of individuals, each of which has some moral worth that we ought to take into account.

6 THE STATE AND THE ENVIRONMENT

This chapter centres on the role of the state but in truth has a wider remit in that it asks what is the most appropriate political vehicle for environmental protection. Alternatives to the state – both from above and below – are considered in an attempt to evaluate the claim made by some that the state remains the ideal institution from a Green perspective. It is noted that much ecological thought tends to have a negative view of the traditional nation state, and advocates instead the devolution of power and authority to smaller political units. This anti-state decentralist position, however, seems seriously at odds with the increasingly global character of environmental problems which necessitates greater co-operation between states and the creation of stronger supranational institutions. More recent Green political thinking, in particular by Eckersley, has sought to rescue the state and, following this recognition, the chapter reviews the literature on the 'Green' state

Greens, decentralisation, and the state

Many green scholars and activists have traditionally had a very negative view of the state. Indeed, they would have regarded the idea of the Green state as 'oxymoronic' (Davidson, 2009: 47). Instead, Greens have traditionally placed considerable emphasis on decentralisation, self-government and egalitarianism as the key operating principles of the sustainable society (see Goldsmith, 1972; Schumacher, 1973; Sale, 1980; Goodin, 1992: 147–56). It is for this reason that it is often said that anarchism is the natural ideological home for Green political theory (see Chapter 9). Perhaps the best-known versions of the decentralised anarchistic model are Kirkpatrick Sale's advocacy of bio-regions, and Murray Bookchin's social ecology position. In the case of the former, self-sufficient and autonomous communities are 'determined by natural rather than human dictates, distinguishable from other areas by attributes of flora, fauna, water, climate, soils and land forms' (Sale, 1984: 168). In the case of the latter, small-scale, self-sufficient communities are necessary because the state is a coercive institution that protects hierarchy. These self-sufficient communities do, however, associate with each other in confederations (Bookchin, 1971).

The idea of small-scale communities, facilitating participatory democracy has had a long shelf-life in Western political thought. It is associated, above all, with the practice of the Greek city-states and the later work of such luminaries as Thomas More, Jean-Jacques Rousseau and G.D.H. Cole, as well as anarchist thinkers such as Godwin and Kroptkin (Pateman, 1970; Eckersley, 1992: 160). In addition, there is a considerable literature on the general desirability and viability of this form of social and political organisation (see Bachrach, 1969). The key question to ask is what contribution can the existence of decentralised, egalitarian and self-governing communities make to the realisation of Green objectives? (see Box 6.1).

Box 6.1 The Environmental Case for Decentralisation

Greens would point, firstly, to the direct environmental benefits of decentralisation. Thus, small, autonomous communities will require self-sufficiency, therefore ending the environmental damage caused by the national and international market in goods and large-scale industrial production. In addition, such communities (particularly in the case of Sale's bio-regions) will foster a greater appreciation of nature's value, beauty and power. Second, there is the argument that a decentralised, self-governing and egalitarian society is one that recognises each person's value as an important and respected member of the community. Each person is entitled to participate equally in the making of laws, is not divided from others through unequally distributed material possessions and feels part of a cohesive community – all characteristics lacking in large and impersonal societies (Martell, 1994: 51–53). As Goldsmith (1972: 51) argues: 'It is probable that only in the small community can a man or woman be an individual. In today's large agglomerations, he is merely an isolate.' In a similar vein, Porritt (1984: 87) states that large-scale industrial societies take 'away our dignity, makes us passive recipients rather than active participants, make ... us dependent rather than self-reliant, alienate ... us from the work we do and the people we live with'.

What we have to consider here is the extent to which the decentralised small-scale community preferred by many radical Greens, whether democratic or not, is the most appropriate means of solving environmental problems. It should be noted that two of the best-known academic accounts of Green political thought, written by authors generally supportive of the radical Green position, doubt whether the small-scale autonomous community is the best model, as does Martell, in his sympathetic commentary on the Green case (Eckersley, 1992: 176–78; Goodin, 1992; Martell, 1994: 54–61; see also Ryle, 1988: 66).

Robert Goodin, in particular, is determined to distance himself from the decentralisation model. He attempts to achieve this by distinguishing between what he calls a 'green theory of value', which seeks to tell us why environmental protection is a desirable objective, and a 'green theory of agency', which is concerned with 'the nature of the mechanisms by which its recommendations (*put simply protecting the environment*) are to be given practical meaning' (Goodin, 1992: 113, emphasis in original). The latter, he argues, is not a central part of the Green case and any particular theory of agency should be accepted or rejected according to the extent to which it helps to promote Green objectives enshrined in the theory of value.

Now, Goodin's primary target here are the various life-style changes and consciousness-raising approaches which Greens are wont to espouse. By distinguishing between value and agency, he is able to recommend turning 'a blind eye' to some of the 'crazier views' or 'green heresies' since these are part of a theory of agency and not value (Goodin, 1992: 16–17). Strategically, of course, Goodin may be correct here but for our purposes it is important to note that he also rejects the decentralisation model on the same grounds: that it does not provide a means of achieving the objectives enshrined in the Green theory of value (*ibid*: 147–66). The central problem, which Goodin and other critics (e.g., Barry, 1999: 86–90) point to, is the inability of autonomous communities to cope with the co-ordination that would be required to deal with environmental problems which, as we have seen, are transnational or even global in scope.

Greens might answer this charge in two ways. In the first place, they might argue that the decentralised communities advocated would be much more environmentally aware and would appreciate the need to act collectively to solve environmental problems. 'Think globally, act locally' is a phrase commonly used by environmentalists to emphasise their dual concern. It is not clear, however, why decentralised communities would be more environmentally aware. It may be the case that people in large urban areas are more environmentally enlightened. Even if it were to prove generally true, there would still be no mechanism whereby the occasional despoiler could be brought into line with other communities. This could prove fatal since those communities making an effort to protect the environment may well conclude that it is not worth their while continuing to pay the cost while other 'free riders' are getting the benefits without contributing. Finally, co-ordination is not just about the exercise of authority. Even if all of the communities were willing to behave responsibly, there would still be a need for a body to co-ordinate their activities, identifying problems and initiating a response.

The second response to the criticism that decentralisation is not appropriate for the achievements of environmental objectives is the point that radical Greens are not opposed to co-ordination, the need for which appears in much of the Green literature (Goodin, 1992: 151–52; Martell, 1994: 56). The problem, however, is they are not particularly coherent. At most, they are prepared to accept that larger co-ordinating bodies should be informal networks rather than permanent unified structures, and certainly nothing like a nation state. Moreover, in so far as they are more coherent (as in the case of Bookchin's model) then they look very much like the state that is rejected in the first place, albeit a more democratised and decentralised version (Barry, 1999: 81). So, there appears to be a strong case for an institutionalised enforcement and co-ordination body not dissimilar to the nation state or even a larger entity, equivalent to, say, the European Union. It is true of course, as Davidson (2009: 62–64) points out, that, as we shall see below, the existing states system suffers from a similar co-ordination problem. At the very least, however, 'reaching agreement may be more difficult under stateless conditions because of the number of parties to the agreement' (Barry, 1999: 90).

The level of criticism directed against the Green emphasis on decentralisation has been pronounced, although some defenders of the decentralisation model still argue their case (Davidson, 2009). In general, though, Green political theorists have increasingly turned their attention towards the role of the state in the international system, on the one hand, and the development of a specifically 'Green' state on the other. Both of these options are explored below.

The state in the international system

The modern politics of the environment is increasingly being fought out at the international level. Indeed, the historical chronology of the environment can be characterised by reference to international gatherings such as those held at Stockholm in 1972, Rio in 1992, Kyoto in 1997 and Copenhagen in 2009 as well as the influential, and impeccably internationalist, Brundtland Report published in 1987 (see Brenton, 1994; Elliot, 1998, Chapter 1). What Britain or any other sovereign state does in environmental terms, then, is now, to a great extent, dependent upon agreements reached at the supranational level. Equally importantly, it is widely recognised that the most critical environmental problems can only be tackled internationally.

The internationalisation of environmentalism

International agreements concerning the environment are not a new phenomenon. It has long been appreciated that pollution is no respecter of national boundaries and requires co-operation between states to deal with it. Equally, animals, and particularly birds, do not respect national boundaries and so wildlife conservation agreements are required to prevent a particular species being protected in one country while being vigorously exploited in another. Further, live animals, and products derived from them, are now an important element of international trade and only supranational action can effectively constrain such trade if it represents a threat to a particular species or involves – as, for instance, does the trade in exotic birds – the infliction of a considerable degree of suffering (Boardman, 1981: 4).

In the last 30 or so years, however, there has been an increasing recognition that environmental problems necessitate supranational solutions and, concomitantly, the study of environmental issues within the academic field of international relations has become much more important too (Vogler, 2008). This international emphasis is partly a product of the general rise of social and political interest in the environment, which, in turn, can be at least partly explained by the severity of environmental problems. Of greater importance is the fact that environmental issues are peculiarly international in scope. It can be readily admitted that some areas of environmental damage do not have a transnational character, but international agreements may still be necessary or desirable in the interests of trade competitiveness. Thus, one obstacle to the development of stringent national environmental laws is the pressure that is likely to be applied by business organisations, which see themselves being economically disadvantaged since they have to pay the costs of environmental protection while their competitors in other countries do not. Thus, one of the major justifications for the European Union's initial intervention in environmental matters was precisely on the grounds of harmonising national laws rather than any specific interest in environmental protection per se (Simpson, 1990: 29).

More often than not, though, traditional environmental problems have had direct supranational consequences. Moreover, the international dimension of environmentalism has become more prominent because of the identification of genuinely global problems, such as climate change and ozone depletion, in the 1980s. Only global co-operation can hope to solve these problems. What were originally national or regional problems have, with the threat to the ozone layer and the climate added to the equation, taken on a global character. Deforestation, for instance, has long been identified as an environmental problem, but whereas the destruction of the rain forests was once a regional concern, affecting the stability of rainfall and river patterns and promoting soil erosion, the recognition that deforestation contributes to climate change transforms it into an issue of global significance. Similarly, the massive increase in road transport has many local, regional and national effects but, as an important source of CO_2 emissions, it is also now inextricably linked with climate change and the international attempts to deal with it.

The environment and approaches to international relations

Perceptions of international environmental regimes are coloured by the general approach to international relations adopted. Traditionally, there have been two main theoretical traditions, but, in recent years, these have been joined by a number of alternatives. It is important to note, initially, that these theories are essentially empirical rather than normative in character. In other words, they seek to tell us how the world is organised rather than how it ought to be organised. Thus, we may wish to advocate the kind of cosmopolitan theory of justice described in Chapter 8 in which the dominance of sovereign states is challenged normatively, while, at the same time, recognising that the realist description of sovereign states in a state of anarchy is more accurate as a description of the international system.

From the perspective of the so-called realist tradition, the dominant theoretical approach in international relations, the difficulties associated with concluding effective environmental agreements will come as no surprise (Morgenthau, 1962; see also Waltz, 1979; Dryzek, 1987, Chapter 6). The realist tradition developed as a challenge to the idealism of much international relations in the wake of the First World War. Realists argue that the natural condition of relations between sovereign states is one of anarchy. The world's political system is equivalent to Hobbes' state of nature with each sovereign state seeking to defend its own self-interest, often defined in economic terms, and being extremely suspicious of others. Given that a world government is unrealistic (Ophuls, 1977), the prospects for effective international co-operation are bleak. This is because, even if the parties stand to benefit equally, without a Leviathan – a body with power over the participants – there can be no guarantee, as in the critique of the Green emphasis on decentralisation, that some 'free riders' will not try to get the benefits without paying the costs (Weale, 1992: 189). Co-operation between states, according to the realist approach, will only occur when there is a dominant state that can impose its will on others.

From the perspective of the realist tradition, then, the central problem of concluding international agreements is that the participants are all sovereign states with no higher legal force standing above them. In environmental terms, the problem is particularly acute because of the dichotomy between the interdependence of the world's ecology (and markets) which recognises no political boundaries, on the one hand, and the fragmented nature of the world's political system consisting of 170 sovereign states on the other (Hurrell and Kingsbury, 1992: 4). The realist school of thought sees no answer to the problem raised by this dichotomy. There is a 'real possibility', Hurrell (1995: 130) writes, 'that our dominant forms of political organization may be inadequate to manage the relationship between humankind and the natural environment on a lasting and sustainable basis'.

If the realist position is an accurate description of international politics then environmental protection on a global scale is going to be very problematic, to say the least. As Paterson (2006: 57) points out, 'the world is for realists "doomed" to unsustainability and crisis'. One beacon of light here is that, from a realist perspective, environmental degradation may be combated because it is regarded as a security issue. Environmental insecurity, it is argued, threatens to become a major source of inter- and intra-state conflict. So, from within the realist tradition, a new branch of the discipline – environmental security – has emerged (Elliot, 2004: 201–22; Swatuk, 2006). Here, it is suggested that environmental problems threaten new sources of violent conflict between and within states (Homer-Dixon, 1991, 1999). Control of scarce resources, in particular water and arable land, is likely, it is argued, to result in conflict, as might the increase in environmental refugees caused by the effects of climate change. In addition, it is undoubtedly the case that military conflict can cause enormous environmental damage, the burning of Kuwait's oilfields during the 1991 Gulf War being a prime example (Barnett, 2001).

The link between environmental scarcity and inter- and intra-state conflict has been challenged. As Elliot (2004: 206) points out, although the state of the environment has played a part in causing conflict in many arenas, particularly in Darfur, Kenya and Haiti, there is 'little compelling evidence that environmental scarcity has been a primary cause of any major sub-national or inter-state conflict'. From a critical perspective, it has been argued that environmental problems in general, and the threat to the environment from military action itself in particular, demands the 'demilitarisation' of security, and the adoption of greater state co-operation (Dalby, 2002).

Without wishing to question the fact that political sovereignty can provide obstacles for the development of international co-operation – or that nation states remain key actors in international environmental politics – the realist perspective does not sit well with the number of international agreements which have been made – there are now about 500 multilateral environmental legal instruments (Vogler, 2005: 2) – and, more to the point, with the fact that at least some of them seem to work reasonably well. This growth of international co-operation gives credence to the utility of the alternative, pluralist or liberal-institutionalist model of international relations (see Keohane, 1989; Keohane and Nye, 1977), which Vogler (2008: 365) describes as 'the predominant IR approach to global environmental change'. This model devalues the importance of nation states and emphasises the co-ordinating role that can be played by other actors such as international organisations, NGOs and epistemic communities in the creation and effective implementation of international regimes. As Hurrell (1995: 137) claims, according to this view, 'power is shifting to institutions above the level of the state, driven by the need to solve common problems in an increasingly interdependent world' (see also Held and McGrew, 1993) (see Box 6.2).

Box 6.2 The Liberal-Institutionalist Model and the Environment

There is some empirical support for the liberal-institutionalist theory, as an explanation for international environmental politics (Hempel, 1996; Rosenau, 1993). The case for the liberal-institutionalist position is based partly on challenging the assumption made by advocates of the realist approach that states act as unitary actors and are not subject to the often-competing demands of domestic actors. This, it is argued, is to ignore the pluralistic nature of Western liberal democracies and the opportunities available for interest groups, and other actors, to influence negotiating positions at a national level (Chasek et al., 2010: 29). Moreover, much is made from this perspective of the important role played by non-state actors at the supranational level. The claims made for pluralism in international politics should, however, be weighed against the weak character of many international environmental agreements, as well as the difficulty of ensuring compliance with the agreements made. It is undoubtedly the case, and liberal institutionalists would not deny this, that state sovereignty remains a central feature of international politics.

Radical political ecology

It is important to point out here too that, from an environmental perspective, the devalued role of the nation state suggested by the liberal institutionalist approach is not necessarily beneficial. Indeed, the continuing degradation of the environment, and the difficulty of concluding effective international treaties, may best be explained, not by the unwillingness of nation states to act, but by the role of powerful, globally organised economic interests and the structure of international capitalism. This is the basis of an alternative radical political ecology approach which, it is argued, has greater explanatory utility than the two traditional models described above (see, in general, Linklater, 1996 and for a specific application to environmental issues see Hurrell, 1995: 141–44; Paterson, 2000, 2007, 2016; Saurin, 1993; Williams, 1996).

The radical political ecology approach, then, represents a challenge to the utility of international co-operation between states. As Carter (2007: 268) observes, 'many sources of environmental degradation may be beyond the reach of environmental diplomacy, because they reside in broader structural factors such as globalisation and the system of international trade'. International environmental agreements therefore achieve very little, being too vague and including too many compromises. Radical political ecologists therefore provide a normative rejection of the states system. As an alternative, new sources of governance – involving, for instance, transnational NGOs and the creation of local political units along anarchistic lines – are advocated (Paterson, 2000: 62).

Paterson (2000, 2007, 2016) provides one of the most comprehensive versions of the radical political ecology approach. He argues that the problem with traditional approaches to international relations is that they assume that environmental degradation occurs by accident, and the debate is simply over whether environmental problems can be solved in the existing states system. For Paterson, a focus on a study of international co-operation is too narrow since it overlooks the contribution that existing interrelated power structures – centring on the state system, capitalism, knowledge and patriarchy – have made to environmental despoilation in the first place. In other words, far from being a potential solution to environmental problems, the existing states system is the main problem. For one thing, the states system is anthropocentric in character in that it is concerned only with the human benefits of environmental protection. A genuinely Green approach to international relations would have to incorporate those characteristics we identified in Chapters 3 and 4, namely an ecocentric ethic, some recognition of the limits to growth, and a radical restructuring of institutions, in order to correct the 'ecological blindness of IR theory' (Eckersley, 2010: 258). Global environmental regimes can then be judged from this perspective, and, indeed, found wanting (Laferriere and Stoett, 1999: 167).

Traditional international relations scholars, for the most part, do not engage with the radical political ecology critique. One exception is Vogler (2005), who seeks to defend the utility of the international environmental co-operation of states. While there may be 'a great deal of depressing evidence' for the radical political ecology thesis, the claim that states cannot act independently of powerful global economic interests is disputed by all but Marxist-inspired scholars. Moreover, Vogler suggests that radical political ecologists underestimate the importance of international co-operation. Not only does it create (sometimes successful) multilateral regulatory frameworks, but it also promotes norms of environmental responsibility, the dissemination of scientific knowledge and capacity-building. Finally, as Paterson (1999) himself points out, such a Green theory of international relations, encapsulating a 'normative rejection of the states-system', is at present underdeveloped, and alternatives to the states system that have been suggested are not very realistic (Vogler, 2005: 249).

Two additional approaches to international environmental co-operationare worth mentioning here. The first focuses on gender. As we will see in Chapter 9, feminists argue that environmental degradation is an issue particularly associated with women, because of their nurturing role that brings them closer to nature, and because, like nature, they can empathise with oppression. It is clear that environmental degradation tends to affect women more than men. This is partly because the poorest are disproportionately likely to be women, and the poorest are hardest hit by environmental degradation. In addition, women's lives are much more bound up in the use of energy, land and water. As a result, it is suggested, the gender inequality clearly apparent in environmental decision-making is a 'roadblock' to sustainable development, and this can only be overcome by greater participation by women (Elliot, 2004: 125–28; Bretherton, 1996).

The second approach is constructivism. A 'pure' or 'thick' constructivism rejects 'the notion of an external reality independent of our knowledge and conceptions of it' (Hay, 2002: 199). A weaker version does not deny the existence of an external reality but seeks to examine 'how states respond to, and how international cooperation is shaped by, the introduction of new information or ideas' (O'Neil, 2009: 11). For a constructivist, interests do not simply emerge as a reflection of objective material circumstances (as realists suggest) but they emerge as the result of the impact of a dominant system of ideas. For constructivists, then, interests are a product of beliefs as much as inevitable material factors. Obviously, this opens up a role in particular for epistemic communities as influential providers of knowledge upon which ideas are based. We saw in Chapter 3, for instance, how the ideology of ecological modernisation has had a significant impact on policy in those states that have accepted it. A crucial implication of this focus on beliefs is that, as Wendt (1999: 134) points out, 'we can have more hope of changing' these belief systems than we can of changing interests derived from the organisation of material forces (see Box 6.3).

The Green state

When discussing the appropriate state form for environmental protection, it is worthwhile starting with, if only to quickly reject, an authoritarian model. An authoritarian strand, particularly prevalent in the

1970s, emphasises the need for drastic coercive solutions – particularly in the areas of reducing consumption and cutting population size – thought necessary to deal with the severity of the environmental crisis (Hardin, 1968; Heilbroner, 1974; Ophuls, 1973). These accounts tend to be based on a Hobbesian model of humans as ultimately selfish and individualistic, requiring an authoritarian state (the Leviathan) to keep order (for a more balanced account of authoritarianism in practice see Beeson, 2010, 2016).

Box 6.3 Constructivism and Ecological Modernisation

One of the few constructivist accounts of international environmental politics has been provided by Bernstein (2001). He argues that a shift in the norms governing how the environment was perceived – from an emphasis on limits to growth in the 1970s to sustainable development within a liberal economic order in the 1980s and 1990s – was reflected in the environmental regimes developed in the respective periods and the attitudes of the states involved in them. For the constructivist, then, it is norms, and not state interests, which is the key explanatory variable. As Bernstein (2001: 10) points out, the institutionalisation of liberal environmentalism 'reflects more than the sum of material interests ... Ideas needed to intervene from some source to create, modify, or, at the least, find a focal point around which existing interests might converge and consensus might form'.

The authoritarian theme is very much the exception in environmental literature – although the degree to which a genuinely Green state must have authoritarian tendencies is a moot point (see below). Rather, the attention of Green political theorists has been, as we have seen, on decentralised political structures. More recently still, there has been a shift back to the state, echoing developments within political science in general (Evans et al., 1985). States, it is argued, still matter. The state 'still remains the primary and most persuasive form of political governance in the world today' (Barry and Eckersley, 2005: xii). States are responsible for implementing and enforcing national environmental regulations, and international treaties and organisations were created by states, and are dependent on states for their continued existence (Duit et al., 2016: 3).

The renewed interest in the state has produced a great deal of empirical work on the Green credentials of existing states. Various labels (ecological state, environmental state, eco-state) have been used to denote states that possess 'a significant set of institutions and practices dedicated to the management of the environment and societal-environmental interactions' (Duit et al., 2016: 4). Clearly, some states fare better than others in this regard. Gough (2016), for instance, identifies four distinct types of environmental governance which he characterises as 'established', 'emerging', 'partial' and 'weak' (see also Christoff, 2005; Dryzek et al., 2003; Gough, 2016).

One empirical account sees the development of the eco-state as following a similar pattern to the emergence of the welfare state in the 20th century (Meadowcroft, 2005). In particular, both involve the expansion of state regulation to new areas of social life in order to rectify market failures. As critics have noted, however, 'crucial parts of the analogy between welfare states and eco-states break down' (Paterson, 2007: 549–51). For one thing, whereas the case for the welfare state increases during periods of economic recession, the evidence suggests that demand for an eco-state is likely, if anything, to decrease. Moreover, welfare states are, arguably, functional for the development of capitalism (in terms, for instance, of creating a more educated and healthy workforce) whereas the relationship between capitalism and eco-states is likely to be conflictual. Finally, the two state forms have different drivers (Duit et al., 2016: 9). Whereas the welfare state is, partly at least, a product of the social democratic constituency in the West, it is by no means clear that environmental protection is a left-wing issue (see Chapter 9). Indeed, as Meadowcroft (2005: 10) himself recognises, it may even be the case that 'the forces that were most significant in building the welfare state represent an obstacle to the ecological state'.

The most concerted attempt to develop a normative model for a Green state has been provided by Eckersley (2004) (see Box 6.4). She recognises that hers is a revisionist account given the distaste for the state that exists in much Green political thought. The justification for bringing the state back in is, at least in part, a product of the pragmatic recognition that the state remains the most important form of political organisation, as the site 'of social and political power' (Eckersley, 2004: 5). Eckersley adopts a version of critical theory (critical political ecology) which, on the one hand, tries to avoid utopian dreaming while, on the other, avoids the conservatism endemic in a mere description of the practices of existing states. The focus on greening the state is therefore part of the process of looking for 'emancipatory opportunities that are immanent in contemporary processes and developments' (Eckersley, 2004: 4).

Box 6.4 Eckersley's Green State

There are a number of elements to Eckersley's Green state. The first is that the state must facilitate a thriving 'green public sphere' in order to create greater ecological consciousness. The state therefore must steer 'society along more ecologically sustainable lines' (Eckersley, 2004: 86). It must also be outward-looking – seeking to co-operate with other states to tackle environmental problems. Moreover, the Green state must more actively challenge property rights (and the neo-liberal economic order) and entrench environmental rights. Most importantly, in the case of the latter, this includes the 'constitutional entrenchment of an independent public authority – such as an environmental defender's office – charged with the responsibility of politically and legally representing public environmental interests, including the interests of future generations and nature' (*ibid*: 243–44). As a result: 'The purpose of the green constitution would be to provide a structure of government that enables, and where necessary enforces, ecological responsibility on behalf of the broader community at risk' (*ibid*: 245).

Above all, the Green state must be impeccably democratic. Not only must it recognise the interests of currently living humans within the state's territory, but it must also recognise the democratic claims of non-human interests, as well as those humans living outside of the state's territory affected by decisions made. It must provide opportunities for meaningful participation and deliberation. Equivalent attention should be paid to the way in which preferences are formed as to the way in which preferences are aggregated. The Green state is, for Eckersley then, synonymous with ecological democracy. She places great store on the capacity of this reformed democratic system – centred on the concept of deliberative democracy – to produce Green outcomes from decision-making.

A number of criticisms can be made about Eckersley's model of the Green state. In the first place, a great deal is made about the pragmatic justification for focusing on the state; that it reflects the state's dominance in the global political order. However, if we accept the liberal-institutionalist model (outlined above) pragmatism might in fact justify focusing on the supranational level instead. A related point is that the radical nature of Eckersley's model does not sit particularly well with her claim that the changes advocated are 'imminent'. It is one thing to say that particular changes to the state in a Green direction are normatively desirable, but it is quite another to say that these changes can be foreseen in existing social and political arrangements. Yes, it is true that states have, to varying degrees, become more environmentally aware, and that there have been numerous multilateral environmental initiatives, some of which have been successful. These developments are far removed, however, from the kind of changes advocated by Eckersley – such as the introduction of deliberative democracy, the entrenchment of environmental rights and the representation of non-human interests.

An additional critique is provided by the radical political ecology model (outlined earlier in the chapter). This model is equally critical of the state as it is of the international system of states. That is, Eckersley is wrong to see the state as separate from, and a solution to, the environmental problems caused by the capitalist system. Rather, it is an integral part of the economic structure which has caused the environmental problems in the first place (Paterson, 2007: 549). It follows therefore that 'the state is not only unnecessary from a green point of view, it is positively undesirable' (Paterson, 2000: 62). Criticisms of this position echo those of the radical political ecology critique of the international states system. In particular, much of the case is, at least partly, dependent upon the claim that the state cannot act independently of ecologically damaging economic forces. This neo-Marxist assumption is, for many, open to challenge (Eckersley, 2004: 91–92; see also Vogler, 2005: 235–36). Moreover, irrespective of its desirability, how realistic is it to advocate the complete overthrow of the state and the establishment of smaller anarchistic-style political communities? (Barry and Eckersley, 2005: 258–59). This, of course, takes us back to square one, and the point at which Eckersley seeks to bring back the state in a Green form.

It was pointed out above that, since the 1970s, Green thinkers have rejected an authoritarian state model. However, an authoritarian critique has been made of the kind of Green state advanced by Eckersley. The claim here is that, irrespective of the denials of Green scholars, *all* states which seek to promote environmental protection as their major goal must have authoritarian tendencies. Warren (2006: 379) even wants to go as far as to say that Eckersley's model of the Green state is tantamount to a 'green police state'. Even if this condemnation is clearly exaggerated, it is possible to note a problem with Eckersley's Green state. Her account is impeccably democratic, but the question that has to be asked is what happens if any particular democratic verdict is anathema from a Green perspective? Would the Green state, perhaps in the form of the environmental defender's office, then have to impose on society the preferred Green option?

There is some doubt, then, whether a Green state can be a liberal state. Indeed, Eckersley herself recognises the tension between environmentalism and liberalism, seeking to contrast a liberal democracy with her preferred ecological democracy. This theme is examined further in the context of liberal theories of justice (in Chapter 8) and in terms of liberalism as an ideology (in Chapter 9). For now, note that a key feature of liberal political thought is the notion that individuals should be free to pursue their own conceptions of the good. Some may regard environmental protection, and even the intrinsic value of nature, as an important value, but this is merely one value that must compete with others, some of which – an interest in powerful cars for instance – may actually damage the environment. The key point is that the liberal state must remain neutral and not favour one conception of the good above others. It is difficult to see, therefore, how Eckersley's Green state can be a liberal state. Indeed, a perfectionist theory which holds that the state should promote a shared vision of the good – associated above all with communitarianism (Sandel, 1998) – would seem a more appropriate vehicle for those, such as Eckersley, who want to prioritise environmental protection.

To be fair to Eckersley, she seeks to deny that her model of the Green state is illiberal. Rather, it seeks to 'criticise the social and communicative context in which … preferences are formed and exercised' (Eckersley, 2004: 96–97). In other words, she adopts a very radical version of democracy which, in particular, utilises a deliberative form which, it is claimed, is likely to produce Green outcomes. In the next chapter I will put this claim to the test. For now, consider a scenario where deliberative democracy does not produce outcomes desired by Greens. What then? It is here, of course, that one may have to choose between a democratic procedure and Green outcomes.

Conclusion

The material highlighted in this chapter is concerned, to paraphrase Goodin, not primarily about a theory of environmental value but about a theory of environmental agency. That is, it is not about the objectives environmentalists want to bring about, but how they are to be achieved and, in particular, what is the best political vehicle to advance them. The Green antipathy to the state was noted and assessed, both in the context of the need for global co-operation, and the traditional Green preference for small-scale, participatory, egalitarian and self-governing communities. As a counterpoint, the more recent attempt by Green political theorists to bring the state back in was reviewed.

While I do not wish to be seen as sitting on the fence, the reader is directed to the conclusion, reached by Hunold and Dryzek (2002), that context plays a large role in deciding how far Greens should prefer the state. That is, in certain circumstances – when, most notably, environmental goals do not undermine the state's economic goals and/or when being associated with environmentalism helps to make the state more legitimate in the eyes of the electorate – it makes sense for environmentalists to 'piggyback their agenda onto the state's' (Hunold and Dryzek, 2002: 22). At other times, by contrast, environmentalists should opt for anti-statist solutions. The fact that, in the early days of environmentalism, environmentalists of any hue had no access to the policy-making process helps, at least partly, to explain the Green movement's preference for a transformation of the state. Those days, however, have long gone (Barry, 1999: 95).

A further question now has to be addressed. Throughout this chapter, the concept of democracy has constantly reared its head, and it appeared particularly in Eckersley's account of the Green state. What has not so far been considered in detail, however, is the extent to which democracy itself is compatible with environmentalism. This is the subject matter of the next chapter.

7 ENVIRONMENTALISM AND DEMOCRACY

It is self-evident to most environmental activists and scholars that the Green state must be democratic. This is not surprising since democracy is now almost universally regarded in a favourable light. However, the relationship between democracy and environmentalism is problematic. Indeed, it has been suggested that democracy is in fact incompatible with the effective protection of the environment. At the very least, the relationship is contingent. The recognition that the relationship is problematic has led Green political theorists to suggest ways in which democratic political systems, as currently constituted, might be reformed. This chapter considers two suggestions. One is the claims made for deliberative democracy. The other, more radical, step is the advocacy of an ecological theory of democracy which seeks to expand the boundaries of the democratic community by incorporating non-human entities.

A contingent relationship?

Following Hayward (1995: 174) the relationship between ecology and democracy can be 'one of indifference … one of incompatibility or … one of necessary connection'. In terms of incompatibility, we have seen that there are those who have argued that the ecological crisis requires authoritarian solutions. This school of thought, as we also saw, was particularly prevalent in the 1970s and was associated with the survivalist mentality that followed the predictions of imminent ecological catastrophe. Even if we accept that democracy is not conducive to the resolution of environmental problems – an assertion which is disputed, in as far as these predictions have proved to be overly pessimistic – it might be suggested that we can now afford the luxury of democracy.

Another, subtly different, ecological critique of democracy is the argument that democracy actually encourages un-ecological policies, as opposed to being unable to deal with pre-existing ecological problems. Democracy, according to this view, 'represents the unleashing of desires, of demands that cannot be satisfied and which have to be authoritatively restrained' (Barry, 1999: 196). This critique only applies, however, if we adopt a narrow definition of democracy as merely a means to the end of utility maximisation, which may, in turn, be as much to do with the nature of acquisitive capitalist societies as with democracy itself (Barry, 1999: 197–98).

Another dimension of this critique of democracy is the suggestion that the resolution of environmental problems requires the will of experts to prevail, and their role can sometimes be undermined or devalued by democratic decisions which pander to the, often misguided, views of the masses and the popular press (Shearman and Wayne-Smith, 2007) (see Box 7.1). While not wishing to deny the importance of experts, it is important to point out that environmental decision-making is concerned with competing interests and values at least as much as it is about technical resolutions to problems. As a consequence, it is important to ensure that all interests are represented in the decision-making arena. Moreover, the employment of maverick scientists by fossil fuel industries illustrates the naiveté of assuming that 'experts' are always entirely divorced from the battle between competing interests.

Box 7.1 The Brent Spar, Democracy and Popular Pressure

One case often quoted as an example of an ecologically dubious action being taken as a result of popular pressure was the fate of the Shell-owned Brent Spar oil loading buoy off Shetland (Dickson and McCulloch, 1996; Bennie, 1998). The company had originally, in 1991, decided to sink the unwanted platform in the North Atlantic but were faced with enormous public pressure, led by Greenpeace, against the move. Shell eventually relented and towed the loading buoy ashore. It is not clear, however, that this was the most environmentally sensible move. For one thing, Greenpeace seriously overestimated the amount of toxic substances aboard the Spar. For another, dismantling such structures can be dangerous. The lesson from the episode is that it is not always the case that the most obvious solution, in the public's eyes, to an environmental problem is necessarily the best from an environmental perspective.

Denying that there is a necessarily antagonistic relationship between democracy and ecology is one thing; demonstrating that there is a necessary connection between the two is quite another. In other words, are there any positive reasons from an environmental perspective why democracy should be favoured? A number of commentators have argued that the relationship is, at most, only a contingent one. In a much-quoted passage, Goodin (1992: 168) states that: 'To advocate democracy is to advocate procedures, to advocate environmentalism is to advocate substantive outcomes: what guarantee can we have that the former procedures will yield the latter sorts of outcomes?' As a corollary: 'To the extent that green outcomes takes precedence over strictly democratic outcomes, it ought to be recognised and acknowledged that democracy is being diluted' (Saward, 1996: 93).

There are a number of reasons why democracy might facilitate environmentally beneficial outcomes. The problem with these types of environmental justifications for democracy, however, is precisely that they are contingent. The operative word is 'might'. If, that is, it can be shown that democracy does not perform the functions claimed for it, as a means to the end of the achievement of environmental objectives, then democracy must be forfeited. We need to ask if there is anything about democracy that is inextricably related to Green thinking, such that the existence of the former is an essential characteristic of the latter? One answer is to suggest that sustainability would always be advocated in a democracy because without human survival no democracy can exist.

As Dobson (1996: 136–39) points out, though, any form of government is dependent on human survival and so 'a sustainable society is as much a generalisable interest for authoritarians as it is for democrats'. In a similar vein, Saward (1996), in a rather half-hearted attempt to amend his earlier denial of the essential link between ecology and democracy, suggests that if adequate health care is seen as a democratic right then in so much as environmental degradation damages health then democracy does necessitate environmental protection. Of course, this argument depends on the, challengeable, inclusion of health care as a key democratic right.

Undoubtedly the most convincing tie between ecology and democracy is the argument that only a democratic society can provide the respect which ecology insists is due to individuals. Like the normative claims concerning the limits to growth we considered in Chapter 3, radical Green thinking is based on the assumption that the preferred social and political organisation must be capable of producing more satisfied and fulfilled individuals in control of their own lives, not only as an end in itself but also as a means of making the sacrifice of material consumption worthwhile. Only a participatory democracy, it is suggested, can perform this function, thereby providing the social and political boundaries of a Green society. This is what Eckersley (1996: 223) is getting at when she writes that: 'authoritarianism is ruled out at the level of green principle (rather than on purely instrumental grounds) in the same way that it is ruled out according to basic liberal principles: it fundamentally infringes the rights of humans to choose their own destiny'.

Democratic reform

Recognising the tenuous relationship between democracy and environmental protection, Green political theorists have suggested a number of responses. In so doing, they have challenged some key assumptions in democratic theory. Traditionally, thinking about democracy has been based on a number of assumptions (Saward, 2001). Three are particularly important in the context of this chapter. The first is that democracy only applies to the political unit of the nation state. The second is that democracy is principally about the aggregation of preferences. That is, it is assumed that a democratic outcome is one that measures accurately the preferences of the electorate in devices such as elections and referendums. The third is that the membership of the political community is settled. Debates about the extent of the democratic community have, it is said, been resolved now that universal suffrage has been achieved.

Of course, one option for Green political theorists is to reject democracy, or seek to limit it. As we saw in Chapter 6, the former was a relatively popular response during the survivalist period in the 1970s, and still has some contemporary resonance (Shearman and Wayne-Smith, 2007). A more nuanced response is to limit, through constitutional means, the power of democratic majoritarianism. So, just as many liberal democracies limit procedural democracy by constitutional provisions which protect human rights, the same could be envisaged for environmental rights (Eckersley, 2004; Ekeli, 2009; Hayward, 2005; Mills and King, 2000). In Eckersley's version, 'certain fundamental rights of non-human species (such as the right to exist) should be incorporated and entrenched alongside fundamental human rights in a constitutional bill of rights to ensure that they are not "bargained away" by a simple majority in Parliament' (Eckersley, 1995: 181).

Deliberative democracy

A more common response now is to advocate tweaking democracy to make it more amenable to the goals of environmentalists. A primary example of this is the advocacy of deliberative democracy. Since the 1980s, democratic theory, and indeed arguably political theory itself, has taken a 'deliberative turn' (Bohman, 1998; Dryzek, 2000: 1). The aim of deliberative democracy has been the offering of a remedy, or at least a partial one, to the perceived weaknesses of traditional democracies – soundbite politics, the dominance of money, poor decision-making, declining participation, increasing moral and political divisions and so forth.

The academic scholarship on deliberative democratic theory is extensive and varied with differences over such key issues as the types of communication to be permitted in deliberative forums, the goal and purpose of deliberation, and the most appropriate site of deliberation. The latter difference is particularly contested, with some deliberative theorists focusing on the benefits of small-scale so-called 'minipublics', such as citizen juries (Elstub, 2014; Fung, 2007), whereas others talk in terms of deliberative political institutions such as legislatures (Steiner et al., 2004) or in terms of a 'deliberative system' within civil society (Benhabib, 1996; Dryzek, 2010; Elstub, 2008). Finally, deliberative participants, too, can be non-specialist members of the public or partisans, whether at an elite level or at the grass roots level.

Despite this variety, it is still possible to elicit a number of key features shared among a vast majority of the exponents of deliberative democracy. Firstly, it is argued by deliberative theorists that democracy ought not to be defined in terms of the aggregation of pre-existing preferences in a vote at elections or in a referendum, or in terms of a reflection of the balance of competing interests within civil society, as the pluralist model has it. Rather, for advocates of deliberative democracy, collective decisions are only legitimate if they are made after reasoned and detailed discussion. Deliberative democratic theorists do think political participation is valuable. But not all participation counts as deliberative. Rather, deliberative democrats are concerned with the quality of decision-making and not merely the numbers involved.

The deliberative process, it is suggested, leads to better decisions in the sense that they are more informed, more effective, more just and therefore more legitimate. This is partly because genuinely deliberative arenas ought to be as inclusive as possible with all points of view and social characteristics represented, and an equal chance to participate offered to all of those who are present. In addition, deliberative theorists insist that self-interest should be put aside, as should strategic behaviour designed to achieve as much as possible of a pre-existing agenda. Instead, mutual respect for, and empathy with, the arguments of others is encouraged. For some (e.g., Estland, 1997), the epistemic function of deliberation – its capacity to reach optimum decisions – is paramount whereas for others (e.g., Gaus, 1997) the intrinsic value of deliberation, its educative function and its closer approximation to political equality, is its most important attribute. From an epistemic point of view, the benefit of deliberation, it is said, is that it increases the pool of information available to the participants, and it permits and improves the detection of factual and logical mistakes in citizens' reasoning about the world.

Finally, 'a central tenet of all deliberative theory' (Chambers, 2003: 318) is that deliberation can change minds and transform opinions. The goal, at least in some – particularly early – accounts of deliberative democracy, is to arrive at decisions that everyone can accept, or at least not reasonably reject. It is seen, therefore, as a useful device to tackle issues which seem to involve intractable moral conflicts (Gutmann and Thompson, 1996). That is not to say that unanimity is a real prospect in most cases, and value pluralism is accepted by most advocates of deliberative democracy as a normatively justified obstacle to consensus (Chambers, 2003: 321; Dryzek, 2010: Chapter 5; Friberg-Fernos and Karlsson, 2014: 100; Gutmann and Thompson, 2004: 26–29; Mackie, 2006: 290). Moreover, as Dryzek (2010: 1–2) has pointed out, an equally positive result of deliberation would be where participants do not change their preferences but decide, after un-coerced reflection, to confirm their initial preferences.

As a result, the aggregation of preferences may well still be necessary as an end-point of a deliberative exercise. However, even if there is still disagreement, collective decisions made after deliberation are regarded as more legitimate than the mere aggregation of preferences, not necessarily or not just because of the decisions made, but because of the deliberative procedure followed which engenders mutual understanding. It involves a sense, that is, that all the views of participants are taken seriously and that everyone tries to empathise with the views of others. For Gutmann and Thompson (1996: 3, 83–85), for instance, deliberation should aim at an 'economy of moral disagreement' in the sense not just that the participants' substantive positions have moved closer together, but that there is a greater acceptance of the terms of difference and disagreement.

Deliberative democracy and the environment

A number of general objections to deliberative democracy can be made (see Box 7.2). It should also be noted that it has also been claimed that there are grounds for thinking that deliberative decision-making might produce outcomes that will benefit environmental protection (Dryzek, 1987, 1990, 2000; Smith, 2003). Deliberative democracy, in other words, has Green appeal. There are a number of reasons for this claim. For one thing, since deliberation requires inclusivity, the views of those who seek greater protection for the environment (including those who advocate an ecocentric ethic) are likely to get a better hearing in an inclusive deliberative environment than in traditional campaigning (where economic interests are likely to have great influence), and this might lead to a shift in views, or at least an 'economy of moral disagreement' among all of the participants (Gutmann and Thompson, 1996: 3, 83–85). In addition, since inclusivity is the goal, there is a justification, too, for including non-human interests in the equation.

Box 7.2 Objections to Deliberative Democracy

A number of general objections to deliberation can be made. The first relates to its practicality. It is unrealistic to expect that anything other than a small minority will be able to engage in genuine deliberation (Walzer, 1999). To create a universal deliberative system involving everyone is impossible because of time, logistical and resource constraints. This would seem to mean that we have a choice. We could opt for a genuinely deliberative system which is politically unequal because only a few will be able to take part. The alternative is to maximise participation to enhance political equality through, for instance, increasing the number of referendums held. This would be done, though, at the expense of deliberation. We would therefore seem to be faced with a choice between deliberation and political equality (Parkinson, 2006). Of course, a response would be to say that deliberative democracy is aspirational. Therefore, it can be accepted that its 'idealized and demanding conditions ... can only ever be approximated (rather than fully realized) in everyday politics' (Eckersley, 2004: 127).

The second objection is that some political theorists have argued that without some fundamental reforms to the balance of power

Continued

among competing interests in contemporary liberal democracies, deliberative democracy merely reproduces existing inequalities, advantaging the already powerful. These inequalities are based, in part, on finance, but also go wider, and include the type of communication (rational and formal debate) valued by advocates of deliberative democracy and the possibility of discourse hegemony (Young, 2000). This would seem to particularly disadvantage social movements such as those concerned with environmental protection and animal rights (Garner, 2016).

Thirdly, some empirical research has found that human beings do not always behave in the way that deliberative democracy theory says that they will (Mutz, 2008). There is, for instance, an inability and/or unwillingness to understand complex social, political and economic issues, and a tendency to follow the dominant majority in group settings, to fit in with the crowd. Finally, the emphasis placed on minimising difference and conflict in deliberative theory has been criticised by some for unrealistically denying the real purpose of politics, which is predicated on the existence of widely diverse interests and values (Mouffe, 2005).

One can see, too, how the empathy facilitated by deliberation might be extended beyond humans to include other species and nature as a whole. Goodin (2003; 2005) provides one model whereby deliberation 'within' (because he envisages an 'internal reflective' mode of deliberation occurring within the minds of individuals as an alternative to an 'external collective' mode) allows for the interests of the excluded and the mute (future generations, nature and animals) to be promoted by being encapsulated or incorporated into the interests of the participants (through 'empathetic imagining'). Eckersley (2004: 116–17) adopts a similar model which she describes as 'enlarged thinking'. Linked to the focus on empathy here is the potential of a shift from a democracy based on speaking to one based on listening. As Dobson (2010: 764) points out, '"giving voice to nature" is less a matter of finding ways of literally making nature speak, and more a question of listening harder to what it already has to say'.

A further factor supporting the claim that deliberation is likely to produce outcomes supportive of the greater protection of the environment is the supposed 'moralizing effect' (Niemeyer, 2004: 347) of deliberation. That is, genuine deliberation involves the advancement of arguments by citizens about what is right, and in the general or public interest, and not about what is in the self-interests of participants. Environmental protection is perceived to be an issue that is in the general interest, not least because it is concerned with protecting public goods – and, ultimately, the survival of all humans. Since the aim of deliberation is to shift the debate from narrow self-interest to generalisable interests and values, it might be argued that it is more likely to be beneficial to the goal of environmental protection. As Eckersley (2004: 117) points out, deliberative democracy makes 'public interest environmental advocacy a virtue rather than an heroic aberration in a world of self-regarding rational actors'.

Finally, it might be argued that deliberation is likely to further environmental protection because of its epistemic value. That is, participants in a deliberative forum are likely to be made more aware of the objective reality of environmental problems – their severity and reach – since there is less chance of distortion by powerful economic interests and the media. Following J.S. Mill, the argument here is that truths are more likely to emerge from open-ended democratic deliberation (Dobson, 1996: 139–41). Of course, there is an element of arrogance in the assumption that the truth will lead to outcomes supportive of greater environmental protection. This might be the case if one imagines that human survival is at stake but, as has been pointed out elsewhere in this book, it is more often the case that the interests and values of environmentalism have to compete with others hostile to particular environmental initiatives.

Ultimately, the assumptions of deliberative theory can only be fully tested in practice by attempts to design ideal deliberative forums as experiments as well as by empirical studies of real-world examples. However, because there are so few examples, deliberative democracy remains an aspiration. The findings from empirical research of such real-world examples have been mixed (Mutz, 2008: 522; Ryfe, 2005: 56). Nevertheless, exponents of deliberation insist that it has the potential to change opinions and generate consensus.

What can be said is that there is plenty of evidence from deliberative case studies, including some focusing on environmental issues, of the impact – in terms of changing opinions – of the full, and unmediated, provision of information (see, for example, Barabas, 2004; Button and Mattson, 1999; Elstub, 2014: 179; Fishkin, 2009: 26–37, 84; Goodin, 2008: 38–63; Kuper, 1997; Luskin et al., 2002; Petts, 2001: 220; Sunstein, 2008: 97). However, this is very different from a transformation of values as a result of debate and discussion. The problem is that this epistemic function can be separated from the deliberation process itself.

That is, one could endeavour to provide comprehensive and balanced information on an issue without requiring those making a decision to then deliberate about it. It is certainly the case that, in so far as minipublics have resulted in preference change, it is difficult to prove, as Elstub (2014: 179) points out, that it has occurred as a result of actual deliberation as opposed to the provision of information. It is also the case that if it can be shown that a 'correct' assessment of comprehensive information is more likely to emerge from a panel of experts, then – if the epistemic function of deliberation is all we are concerned about – we would be duty bound to prefer this process over democratic deliberation (Freeman, 2000: 387; Lafont, 2006: 8–9).

The final point to make here is that deliberative democracy remains anthropocentric or human-centred. It is true that Eckersley (2004: 115), for one, seeks to 'defend deliberative democracy as the best means of practically serving' the claims of a non-anthropocentric theory of democracy. Even Eckersley recognises, however, that deliberation has to be supplemented by constitutional provisions entrenching the interests of non-humans. In a system of deliberative democracy without such limits, therefore, it remains the case that the interests of non-humans are only considered if, and until, enough humans think those interests are worthy of consideration. The key point to note, then, is that the relationship between democracy and environmental protection in an anthropocentric version of democracy, even one of a deliberative bent, remains contingent. Deliberation may mean that non-human interests get a better hearing, but it does not guarantee it (Mills and King, 2000: 139–41).

Ecological democracy

Responding to the difficulties the anthropocentric theory of democracy has in representing the interests of non-humans, Green political theorists have explored the case for a non-anthropocentric version. One objection to this idea might be that there is no alternative to an anthropocentric theory of democracy because only humans can identify and articulate the interests of the non-human world. The extent to which this is the case might be disputed, but – assuming it is correct – a possible response is to distinguish between two different forms of anthropocentrism (Garner, 2017). The first, weak, version of anthropocentrism makes the obvious point that the interests of non-humans must be identified and articulated by humans, and the non-human world must be represented by humans who speak on its behalf.

This weak version of anthropocentrism, however, is different from a second, strong, version of anthropocentrism which holds that only human preferences regarding the non-human world ought to be promoted. That is, the interests of non-humans (animals and non-sentient parts of nature) ought not to be represented independently of human interests. In other words, one can distinguish between an objective account of the interests non-humans have (irrespective of the degree to which they conflict with human interests), and the extent to which humans want these objective interests of non-humans to be promoted. The contingent relationship between democracy and environmental protection depends upon this second, strong version of anthropocentrism. That is, the interests of non-humans will be protected in a strong anthropocentric account of democracy only if enough humans want this to happen.

This strong anthropocentric justification is consistent with the current way in which concerns about the environment are expressed. Thus, groups seeking to protect and defend the non-human environment are ubiquitous in liberal democratic polities, and they stand or fall by their ability to persuade other humans to join their cause, and decision-makers to adopt their demands. What should be noticed about the strong anthropocentric perspective, then, is that the interests of non-humans are represented indirectly. That is, the degree of consideration given to the interests of the non-human world depends upon the extent to which humans are concerned about defending and promoting those interests. By contrast, the weak form of anthropocentrism does not rule out the prospect of providing a normative justification for the democratic representation of non-humans irrespective of the wishes of any particular human electorate.

Along these lines, Green political theorists have flirted with the idea of a non-anthropocentric theory of ecological democracy. This involves challenging one of the central assumptions made in traditional thinking about democracy (Saward, 2001). This is that the membership of the political community is settled. Debates about the extent of the democratic political community have, it is said, been resolved now that universal suffrage has been achieved. A non-anthropocentric theory of democracy, however, challenges that assumption. Membership of the democratic community ought to be expanded to include both humans and non-human constituencies, because restricting the political community to humans has the effect of excluding other entities that have important interests and who are entitled to have these interests represented.

Expanding the democratic boundary does not require us to adopt a non-anthropocentric democratic theory if we only want to extend it to include future generations of humans, although even this step would expand the boundaries of the democratic community in a radical way. In theory and practice, the democratic community has traditionally only included currently living humans, and yet it is undoubtedly the case that future generations of humans have important interests which may well conflict with ours. This is particularly the case, of course, with environmental protection. For example, we may want to maximise our economic prosperity but this may cause irreversible problems for generations to come. We might use up all of a particular resource so there is none left for those not yet born, or we might pollute the planet so that the quality of life for future generations will be much reduced or even non-existent.

The case for including future generations in the democratic community is tied up with debates about our moral obligations to them (see Chapters 4 and 8). Ostensibly, it would appear that the case for including future generations presents fewer problems, both theoretically and practically, precisely because it remains an anthropocentric approach. Indeed, a number of countries have explored the possibility of constitutionally enshrining the interests of future generations (Tremmel, 2006).

Even more radical is the claim that democracy is illegitimately anthropocentric. That is, it is too focused on humans. After all, non-humans – animals and nature as a whole – have interests which may conflict with ours. And these interests may be ignored if only humans are members of the democratic political community. As a result, Green political theorists have sought to work out how we might incorporate the interests of the currently excluded (see below).

The all-affected principle

Merely stating that future generations and nature have interests, of course, does not, by itself, justify extending the boundary of the demos to include them. It may be, of course, that the strong anthropocentric version of democracy provides an entirely legitimate account of the place of non-humans within the political process; that there is not a case for going further to establish the democratic right of non-human parts of nature to direct representation. However, many Green political theorists regard the strong anthropocentric version of democracy as normatively deficient. This position tends to be justified on the grounds that it is inconsistent with the all-affected principle – a principle that is widely thought to be the most appropriate means of representing interests comprehensively (see Box 7.3).

Box 7.3 The All-Affected Principle

The all-affected principle is one answer to the so-called 'boundary problem' or problem of inclusion in democratic theory which asks: how do we decide who is to be included in the demos? Who is to count as 'the people'? The all-affected principle responds by saying that 'everyone affected by decisions made has a categorical right to participate in the process of making those decisions'. Many political theorists (and not just Green political theorists) have seen the all-affected principle as the best solution to the boundary problem. As Goodin (2007: 47–50) notes, the all-affected principle provides us with an underlying principle that justifies both democratic decision-making and a democratic means of deciding who is to be included in the demos. For Dahl (1979: 19) likewise, the all-affected principle is valid because the exercise of power over those who have no say in the outcome amounts to the exercise of arbitrary power which infringes the fundamental democratic value of political equality. The all-affected principle is therefore implicit in the central democratic doctrine of government by the governed.

From an environmental perspective, there are two distinct advantages of invoking the all-affected principle. In the first place, it provides a justification for the inclusion of non-humans within the democratic community. Since non-human parts of nature are also affected, often detrimentally, by political decisions, it follows, according to the all-affected principle, that they should therefore have a say in those decisions. Secondly, the all-affected principle accounts for the supranational nature of environmental problems rather better than the traditional democratic domain of the nation state. The increasing dangers of ecological degradation offer, perhaps, the best example of the lack of fit between modern democracies centred on the nation state and the structure of power and interests in the real world. Adopting the all-affected principle would mean that those in one political jurisdiction likely to be affected by environmental degradation caused by the actions of another ought to have a say in the decision authorising those actions.

The all-affected principle therefore supports the case for a cosmopolitan theory of democracy. Here it is suggested (for example by Held, 2006: Chapter 11) that given that citizens of nation-states are increasingly affected, if not dominated, by forces beyond the boundaries of the particular nation state within which they live, then what matters now is ensuring that global forces are controlled by democratic means. This requires that the international level of inclusion be at the level of the individual and not, or not exclusively, the state. As Weale (2007: 239) points out: 'Just as democracy had to make the transition from the city-state to the nation-state, so it must now make the transition to the international global order.'

Assessing the all-affected principle

In an assessment of the all-affected principle, it is important to distinguish between general and more specific objections. One particularly pertinent general objection is the so-called incoherence objection (Goodin, 2007: 52). This is the argument that it is impossible to determine who is to be included in the demos before a decision is made because we will not know until then who has been affected. As Owen (2012: 134) points out, however, it is possible to refine the all-affected principle in a way that deals with this objection. Inclusion in the demos becomes restricted, in this refined version, to all 'whose interests are actually affected by a choice between all plausible possible options open to the collective decision-making body' (see also Goodin, 2007: 56–59 for a similar response to the incoherence objection). This may, of course, have the effect of widening the scope of the all-affected principle somewhat – which might be regarded, in itself, as a further objection (see below).

We can also challenge the more specific claim that the all-affected principle justifies the inclusion of non-humans as members of the democratic political community. In the first place, it might be argued that non-humans cannot be included as beneficiaries of the all-affected principle because they cannot represent themselves. In practical terms, it might be argued that we cannot know what the interests of non-humans are because they cannot tell us, and, what is more, they cannot be represented because there is no one to actually physically do it (Farrelly, 2005: 204).

Neither of these specific objections is insurmountable. As we saw earlier, it is not inconceivable to imagine us being able to 'surmise' the interests of the non-human world, even if it cannot tell us (Goodin, 2003: 215). These interests can then be promoted, by humans with a fiduciary responsibility or duty towards nature, whether or not there is any wider immediate public (human) support for these interests, and whether or not they clash with important human interests. Indeed, political theorists have developed sometimes ingenious schemes in which humans act as proxies for excluded, so-called 'mute', interests such as future generations and nature whose representatives are allocated some seats in legislative assemblies and counter-majoritarian devices are suggested to impede the will of the majority. (Examples of the former are Dobson, 1998; Ekeli, 2005; and Mills, 1996. Examples of the latter are Ekeli, 2009; Eckersley, 1999; Hayward, 2005; Kates, 2013; Thompson, 2010.)

A more challenging justification for denying that the all-affected principle allows for the inclusion of non-humans comes from those who emphasise the significance of agency. Saunders (2011: 286), for example, argues that those lacking agency cannot be enfranchised because, even though they have interests, they are not capable of participating in the demos. As he points out: 'It is bizarre to suggest that animals or unborn children ought to be granted the vote, though they cannot possibly exercise it.' Democracy is a matter of rule by the people, and since those without agency cannot, in any sense, rule, they cannot be included in the demos. Conversely, to exclude those without agency from the demos cannot be an undemocratic move.

It is noticeable that most pre-twentieth century democratic theory – including Locke, Rousseau and Mill – and practice regarded inclusion in the political community to be contingent on cognitive competence. Even notable contemporary political theorists want to impose restrictions on the grounds of cognitive capacity (Dahl, 1989: 122–25; Habermas, 1996: 127). Dahl (1989: 83–93), for instance, argues that democracy is justified on the grounds that self-governing preserves the autonomy of persons whereas to limit a person's opportunity to live under laws of her own choosing is to limit her autonomy. On these grounds, not only would non-sentient parts of nature be excluded from membership of the political community, but sentient non-human animals would be similarly denied entry, because, according to an extrapolation of Dahl's account, of their perceived lack of autonomy (Cochrane, 2009) or, at least, their lack of a particular, higher, kind of autonomy (Frey, 1987).

In answer to the competence objection, it should be noted, firstly, that Saunders is prepared to concede that if one accepts the all-affected principle then the case for enfranchising non-humans is unanswerable (Saunders, 2011: 288). It is necessary, therefore, to tackle his opposition to the all-affected principle by challenging his claim that those lacking agency cannot be enfranchised. Here, it can be argued that he has an unduly narrow view of what being enfranchised means, and thereby gives the concept of agency a significance that is unwarranted. It is true, of course, that, because they lack agency, non-human entities cannot participate in collective decision-making. But that does not, as was indicated above, stop their interests being represented in a democratic political system through proxies, as suggested by Dobson (1998), or by institutions designed for the purpose.

Here, one should not confuse the capacity to participate in decision-making, and the ability to hold representatives to account, on the one hand, with the existence of clearly identifiable interests that could be represented. Children (below a certain age at least), of course, must be disqualified on the grounds of competence from being active members of the demos. But their interests are still affected by decisions made. It is not difficult, therefore, to envisage others elected to the task of speaking on their behalf. Similarly, the severely cognitively impaired lack the competence to participate in the demos but clearly have interests that can be affected by collective decisions. And, of course, if proxies can be elected to represent the interests of children and the cognitively impaired, there is no reason why they cannot be elected to represent the interests of non-humans who are equally affected by decisions made.

Perhaps the most significant objection to the all-affected principle is that its coverage is likely to be extremely, and perhaps unsustainably, wide. Of course, this width is precisely what allows for the inclusion of non-humans in the first place, but logic would seem to suggest that it should go further. For, if entry to the democratic arena does not depend upon the possession of some necessary capacities (rationality, agency, competence and so on) then it logically follows that everything – animate or inanimate – whose interests are affected by a collective decision would be entitled to have these interests represented. This would clearly expand the democratic arena even further to the point where it becomes intuitively difficult to justify.

To avoid the intuitively problematic reach of the all-affected principle, we could limit the scope of the all-affected principle by introducing sentience as a qualifying criteria. So, only entities which are sentient, with the capacity to be aware of the effect collective decisions have on their interests, should be included in the democratic community (Garner, 2017). Such a move may be normatively correct, but, of course, it would – in an echo of the environmental ethics debate considered in Chapter 4 – have the effect of excluding non-sentient parts of nature. In addition, it would render problematic the inclusion of future generations since, although they will be sentient, they cannot be said to be so until they exist.

Some Green political theorists would reject the sentience qualifying criterion and accept the conclusion that the all-affected principle provides a justification for the enfranchisement of non-sentient parts of nature. The view that non-sentient parts of nature do not have interests that can be represented can certainly be questioned by adopting a very broad definition of interests. Ball (2006: 137), for instance, defines interests such that 'x is in A's interest if x is necessary for and/or conducive to A's functioning or flourishing'. The possessor of an interest in this sense does not require sentience and such an interest does not need to be articulated by its possessor. Eckersley (2004: 111) adopts this approach when she states that 'all those potentially affected by ecological risks should have some meaningful opportunity to participate or to otherwise be represented in the making of the policies or decisions which generate such risks'.

Conclusion

We have seen in this chapter that there is a problematic relationship between environmentalism as a value, or preferred outcome, and democracy as a procedure, or theory of agency. There is no guarantee, therefore, that democratic decision-making will result in outcomes desired by those concerned about environmental protection. At best, then, the relationship is contingent. Of course, it might be hoped that currently living humans will, at the very least, recognise that their interests are served by acting so as to ensure that the natural world is protected. Too often, however, this goal conflicts with short-term gains, particularly if the negative consequences are unlikely to be seen until sometime in the future. Green political theorists have suggested a number of ways of dealing with the problematic relationship between democracy and environmental protection. One is to reject democracy in all or part, another is to seek to tweak democracy in order to make it more amenable to environmental protection, and yet another is to reject, not democracy as such, but only anthropocentric versions.

Two concluding points are worth making. One is that even non-anthropocentric versions of democracy will have a contingent relationship with environmental protection. The fact that the interests of non-humans will be represented does not, of course, ensure that those interests will prevail. Following on from this it is important to recognise that the exclusion of non-humans from the democratic community does not mean that we have no moral obligations to the non-human world. Indeed, ultimately, the case for protecting the natural world – and the interests of future generations – depends upon the extent to which currently living humans regard this as the morally right thing to do.

8 THE ENVIRONMENT AND THE IDEA OF GLOBAL JUSTICE

The relationship between environmental sustainability and a concept that is central to politics, justice, has become an increasingly important preoccupation of environmental political theorists (Dobson, 1998; Schlosberg, 2007, 2013; Wenz, 1988). The question here is the degree to which protecting the environment can be regarded as just. The concept of justice has been a key concern for political philosophers dating back at least to the Greeks. The mainstream definition is that justice is about distributing benefits and burdens according to what recipients are due (Campbell, 1988: 19), but there are also alternative, or additional, models based on recognition, participation and capabilities. Defining the concept is the easy task. It is a much more difficult, not to say contentious, matter to decide what these benefits and burdens are, how they are to be distributed, to whom and by whom. The advantage of securing the protection of the environment as a matter of justice is that it represents a strong form of entitlement, to be distinguished from charity or benevolence. Conventionally, in liberal democracies, justice will be protected through constitutionally entrenched provisions which cannot be undone by majoritarianism (Hayward, 2005).

Traditionally, political theorists had been mainly preoccupied with working out principles of justice to be applied in one country. Moreover, the beneficiaries, or recipients, of these principles, particularly for contractarian political theorists, have been exclusively currently living human beings. The environmental debate has challenged both of those assumptions. The fact that environmental degradation does not respect national borders, and that its impact is unequally distributed in the world, has, along with the phenomenon of globalisation, led political philosophers to unlock their national shackles in order to develop so-called cosmopolitan theories which seek to work out global principles of justice (Caney, 2016). Not surprisingly, perhaps, much of the work in the environmental field has focused on the relationship between anthropogenic climate change and justice.

Environmental issues have also raised the issue of expanding the range of actors who have to be considered as recipients of justice. In the first place, since the decisions we take on the environment now will have a long-term impact, and may indeed be irreversible, we have to consider whether the interests of future generations of humans should be taken into account. What, if anything, are they due from a theory of justice? Moreover, if we adopt a non-anthropocentric ethic it is not just human interests – whether those of currently living or those yet to be born – that have to be taken into account in a theory of justice, but also animals (Garner, 2013; Nussbaum, 2006) and even nature as a whole, including sentient non-human animals and living but non-sentient parts of nature (Baxter, 2005; Schlosberg, 2007).

If it can be established that environmental degradation is unjust because it damages the interests of future generations and nature as a whole, then there would be a strong claim for saying that causing damage to the environment is unjust. That is, it is just to act so as to ensure that future generations and non-human parts of the environment survive and prosper. This may, of course, mean that current humans may have to make sacrifices, and their interests would have to be taken account of too before we could reach a verdict. Conversely, if it is concluded that justice is a concept only applicable to currently living humans then it is by no means clear that justice *is* served by protecting the environment. For example, the potential mismatch between justice and sustainability is particularly marked in the 'survivalist' literature of the 1970s. The kind of authoritarian solutions recommended in this literature would undoubtedly trample over what most would regard as fundamental human rights.

Justice, political theory and the environment

The conventional model of justice adopts a distributive paradigm. That is, the concept is concerned with developing a theory justifying a particular distribution of benefits and burdens. Of particular importance here is the work of John Rawls, the Harvard scholar whose book *A Theory of Justice* (1971) is regarded as *the* most important work of political philosophy in recent decades (see Box 8.1). Rawls's conclusion, that only currently living humans are beneficiaries of a theory of justice, derives from his adoption of a contractarian approach. This is an approach to political theory, dating at least as far back as the thought of Thomas Hobbes and John Locke in the 17th century, where principles of justice emerge from an imaginary social contract between rational agents who meet together to decide on the principles of justice under which they are to live. Those not capable of participating in the drawing up of such a contract – those, for Rawls, who are not moral agents – are therefore excluded.

Box 8.1 Rawls, Justice and the Environment

Rawls is important in the context of the environment because he argues that only moral agents - those, firstly, capable of having a conception of their own good in terms of a rational plan of life, and, secondly, a capacity to understand justice and what it requires of them and what they can expect from others - are entitled to be beneficiaries of justice. This excludes future generations, non-human animals and non-human parts of nature from being considered as recipients of justice. This is not to say that they are owed nothing. They may, for instance, be the beneficiaries of moral obligations on the part of moral agents. What is for sure, though, is that they are not due justice. Justice is, for Rawls, therefore, an anthropocentric concept, and a limited one at that by virtue of the fact that only currently existing human are entitled to be beneficiaries of justice.

One response to contractarian theories is to seek to adapt them in order to include the entities left out. One way of adapting Rawls's theory is to adjust his methodology slightly. Rawls imagines an original position whereby the contractors meet. In this original position they are behind a 'veil of ignorance' whereby they have no knowledge of their social position – their income, gender, race and so on – in real society. One possible adaptation is to 'thicken the veil' so as to exclude knowledge of when participants exist – now or in the future – and species. If this is done, it is argued, the contractors will seek to protect the interests of future generations and nature in case they end up turning out to be someone yet to be born, a non-human animal or other non-human part of nature once the veil is lifted. This adaptation tends to be more convincing when it is applied to non-human mammals (Rowlands, 2009), not least because it is easier for the contractors to imagine what it is like to be, say, a chimpanzee, than a living, but not sentient, part of nature.

It is a central feature of liberal contractarian theories that justice involves creating a society in which people are free to pursue their own conceptions of the good: a set of beliefs that shapes the way they live.

Principles of justice are then devised to facilitate individuals living in the way that they see fit. Clearly, if a failure to protect the environment was regarded as a threat to the free pursuit of conceptions of the good then there would be a case for including it as a constitutional essential. It is more likely, though, that a desire to protect the environment, and a belief in the intrinsic value of nature, will be regarded as a conception of the good to stand alongside other, sometimes competing, conceptions of the good. This is clearly inadequate from the perspective of environmentalism (Dobson, 1998: 197–98) since some may wish to pursue activities that damaged the environment, and the state, under a liberal theory of justice, would not be justified in intervening to prohibit such activities.

Arguably, all is not lost if environmental protection is excluded as a principle of contractarian justice. Contractarian political theorists, such as Rawls and the British political philosopher Brian Barry (1995), suggest that environmental protection should be the subject of political debate, and democratic votes, even though it is not a matter of justice. Since constitutionally entrenched provisions protecting the environment do not yet exist anywhere, this has the advantage of reflecting the political reality of environmental politics. As Baxter (2000: 49) points out, this means that 'whether or not any component of non-human nature is saved depends on whether or not enough people can be persuaded' that they ought to be. Rawls's and Barry's liberalism is, in Bell's words (2002: 721), then, a '"contingently green liberalism" rather than an "intrinsically green liberalism"' and it is 'up to the environmentalist to persuade enough of their fellow citizens of the value of their "green ideals"'.

Another response to contractarian theories of justice is to reject the claim that environmental protection is not a matter for justice. This involves rejecting contractarian theories in favour of ideas more conducive to a positive link between justice and environmental protection. One possibility is to adopt a communitarian theory of justice whereby the state promotes the protection of the environment as a goal. We might, alternatively, employ arguments designed to support an ecocentric ethic – based on sentience or the capacity to live – as important justifications for applying the label justice to the protection of nature. To do so is to break the link between moral agency and justice and to distinguish between distributors and recipients of justice (Dobson, 1998: 64–69). Since we discussed these ideas in Chapters 4 and 5, the rest of this chapter will be concerned with seeking to examine the case for an anthropocentric theory of environmental justice.

Although it has dominated in academic writings on justice, the distributive paradigm is not the only model available to adopt. Schlosberg (2004, 2007, 2013), in particular, utilises a variety of approaches to develop a pluralistic theory of environmental justice which combines 'numerous concerns into a broad and multi-faceted approach to justice' (Schlosberg, 2007: 11) (see Box 8.2). In the first place, justice can be served by facilitating participation or a 'seat at the table' (Schlosberg, 2013: 40). Sterba (1998: 142), for instance – in echoes of the all-affected principle that was discussed in Chapter 7 – argues for a principle of procedural justice where 'everyone, especially minorities, should participate in the selection of environmental policies that affect them'.

Relatedly, one can also invoke the idea of justice as recognition particularly associated with Iris Young (1990). This approach is based on the assertion that it is important to examine the underlying causes of the maldistribution of resources (and specifically, in the context of this chapter, environmental goods and bads). Here, a lack of recognition and respect in the political realm 'demonstrated by various forms of insults, degradation, and devaluation' (Schlosberg, 2007: 14) holds the key. In other words, a failure to show respect is a harm that is every bit an injustice as an unjust distribution of resources. It might be argued that recognition is not a distinct theory of justice, but a part of the distributive paradigm in the sense that a just distribution assumes recognition and respect (Schlosberg, 2007: 20). Nevertheless, an unjust distribution might be a product of a lack of recognition and respect, such as the devaluing of the lives of members of minority communities who are forced to suffer the effects of environmental degradation (see below).

Also applicable is the capabilities model of justice developed, in particular, by Sen (2009) and Nussbaum (2004; 2006; 2011; see also Schlosberg, 2012 and Holland, 2014 for applications to environmentalism in general). The emphasis on capabilities originates from the work of Amartya Sen, who developed the idea as an alternative to the dominant model of resource distribution (Sen, 1993). That is, rather than focusing on, say, the distribution of income and wealth, the state should instead ascertain the degree to which a particular resource distribution allows individuals to flourish. This, in turn, requires the identification of basic capabilities that are required for individuals to lead a dignified life.

Nussbaum has developed and popularised Sen's approach, identifying a list of capabilities initially to be applied to humans only, but then extended to encompass non-human animals too. These capabilities are required in order to enable humans, and animals, to flourish, and ought to be upheld at least to a threshold level. For Nussbaum (2006: 75), 'a society that does not guarantee these to all its citizens, at some appropriate threshold level, falls short of being a fully just society'. Nussbaum adopts the same capabilities for animals as she does for humans, although she recognises that not all of the capabilities will apply to all non-human sentient species. These capabilities are (Nussbaum, 2004: 315–17): (1) life, (2) bodily health, (3) bodily integrity, (4) senses, imagination and thought, (5) emotions, (6) practical reason, (7) affiliation with others, (8) a meaningful relationship with other species and with nature, (9) other species (10) play, and (11) control over one's environment.

Box 8.2 Schlosberg's Theory of Environmental Justice

Schlosberg does not disagree with the traditional distributional model but argues that it does not encapsulate all that can be said about justice, and, what is more, does not exhaust what environmental justice movements understand by the concept. As he points out (Schlosberg, 2013: 50): 'Many attempts to define environmental or climate justice have been too detached from the actual demands of social movements that use the idea as an organising theme or identity.' For example, the capabilities model helps to identify the distributional model that will maximise flourishing, the most equitable distribution requires recognition, and 'demands for more broad and authentic public participation is often seen as the tool to achieve both distributional equity and political recognition' (Schlosberg, 2007: 26).

In addition, Schlosberg argues, this pluralistic approach is more able, than the distributive model alone, to incorporate the non-human realm into a theory of justice. We saw in Chapter 5, for instance, that Nussbaum wants to apply the capabilities approach to incorporate non-human animals within a theory of justice. Schlosberg (2013: 44) wants to take this further by arguing that an interruption of the capabilities and functioning of living systems is an injustice not only to human beings and non-human animals but also to 'all of those non-humans that depend on the integrity of the system for their own functioning'. Of course, there is potential for conflict between the functioning of humans and non-humans (Cripps, 2010), but then, as Schlosberg (2013: 45) notes, conflict issues are prevalent in any theory of justice (including anthropocentric ones) and have to be worked through accordingly.

Intra-generational and inter-generational justice and the environment

Questions relating to intra-generational justice arise when environmental problems have a disproportionately negative effect on a group or groups in a particular society at a particular time. Initially, concern was limited to the localised impact of the distribution of environmental benefits and burdens. That is, environmental 'bads' (such as pollution) were distributed so as to negatively affect disadvantaged communities and these same communities were denied equal access to environmental goods (such as green spaces, public transport and fresh food). For some authors (see, for example, Pellow, 2004) environmental injustice was suffered, in particular, by black communities.

This recognition of disadvantage led, particularly in the United States, to the emergence of grassroots environmental justice movements, designed to draw attention to and eradicate a perceived injustice (Bullard, 2000; Newell, 2006). A key development in the United States was the creation of the Love Canal Homeowners' Association in the late 1970s to seek compensation when it was discovered that the community, in New York state, had been built on a toxic waste dump. The determined leadership of Lois Gibbs resulted in national media coverage and the intervention of President Carter, who ordered that the people be relocated (Dryzek, 2013: 213–14). Today, numerous environmental justice campaigns exist focusing, in particular, on the location of hazardous waste, and national organisations, such as the National Black Environmental Justice Network in the United States and the Black Environmental Network in the UK, have been formed.

The concept of environmental justice is now applied to a wider range of issues and the national focus of environmental justice movements has been dwarfed by the question of justice on a global scale (Carmin and Agyeman, 2011). In this so-called second generation of environmental justice, attention was altered from a demand for 'not in my backyard' to 'not in anyone's backyard' (Vanderheiden, 2016: 323). The growing interconnectedness in the world of peoples and sovereign states has provided a fillip for extending justice beyond national boundaries. Limiting a discussion of justice to the internal affairs of wealthy Western states seems trivial, given the staggering inequalities between different parts of the world. As a reflection of this wider concern of the environmental justice movement, climate change is seen, above all, as an issue that has a disproportionate impact on disadvantaged groups.

Recognising the global character of justice, political theorists have sought to develop theories of justice that are global in scope. There is little agreement, however, on what our moral obligations should be to those outsiders who do not belong to our community. For some philosophers, the moral obligation on the richest to help the poorest is clear cut. For some, the sacrifices required will make many of us feel uncomfortable. At one end of the spectrum is the position represented by Peter Singer (1985) (see Box 8.3). Singer's cosmopolitan ethic is echoed by others who argue that principles of justice apply to global society (see, for instance, Beitz, 1979; Pogge, 2002). This contrasts with others who, while not denying that we have humanitarian duties to relieve extreme poverty, argue that justice only applies to relations we have with our fellow countrymen (Nagel, 2005). All cosmopolitan theories of justice, of course, are vulnerable to the charge, often put forward by exponents of so-called non-ideal theory, that they are putting forward principles which are unrealisable.

Box 8.3 Singer and the Demands of Global Justice

Singer offers an uncompromising account of our obligations to those in poverty throughout the world. Adopting a universal utilitarian ethics, he argues that if we can help towards eradicating global poverty then we ought to. To justify this position, he uses an analogy. Imagine that we are walking past a child drowning in a shallow pond. Would it be morally obligatory to rescue the child given that the costs to us are negligible whereas our failure to do so will be catastrophic for the child? Most would answer yes, but of course if we extrapolate to the needs of the world's poor exactly the same equation exists. Singer wants to go further though. We should, he argues, give all of our income to alleviate Third World poverty once our own necessities of life are met (Singer, 2002), and such a gesture is a moral duty and not a matter of charity.

A key question for this chapter is the degree to which environmental protection is consistent with global social justice. In some ways, firstly, it should be noted that environmental issues are inextricably related to development issues, since the biggest threat to the global environment comes from economic activities in the developing world (Miller, 1995). A major concern of the consequences of economic growth in the developing world is a massive increase in greenhouse gases; another is the depletion of non-renewable resources. A further fear is that the increase in population and urbanisation will put stresses on the food production sector and that the threat of shortages will encourage ever more intensive agricultural practices, causing further environmental damage.

Under threat too are some of the most species-rich regions on the planet, and just as increases in pollution have consequences for the developed world, so does a decline in biodiversity (see Chapter 2). The vast majority of the world's most species-rich habitats can be found in developing countries. Over half of all the species so far identified, for example, can be found in the tropical rainforests (Elliot, 2004: 30). By contrast, only about 3% of the global total of recorded species can be found in the UK (Pearce et al., 1993: 102). The active involvement of developing countries in the protection of biodiversity is arguably a more urgent task than, say, action to reduce greenhouse gas emissions. In the latter case, developed countries can take action to reduce their own CO_2 emissions while seeking the long-term participation of developing countries as their economic development proceeds. No such luxury is present in the case of biodiversity, where immediate and irreversible damage is being done. If biodiversity is to be protected effectively, therefore, the immediate and active involvement of developing countries is required.

On the other hand, social justice would seem to be served by environmental protection in the sense that the national picture of disproportionate impact is echoed at the global level. Thus, it is the poorest countries in the world that are most affected by environmental damage and they are least able to adapt to it. This applies particularly in the case of climate change because the impact of droughts and rising sea levels will be felt most by the poorest people in Africa, Asia and the Pacific. Moreover, most of these environmental problems have been caused by the developed world. Elliot (2004: 3) reveals the stark reality when she writes that: 'The richest 20 per cent of the world's population … accounts for 86 per cent of total private consumption expenditure and consumes 58 per cent of the world's energy whereas the world's poorest 20 per cent consume less than 5 per cent of global resources.'

Social injustice contributes to environmental degradation in the sense that it encourages practices which cause environmental damage. Indeed, there is a school of thought which argues that environmental degradation in developing countries is in large part a product of their dependence on the West (the following account of this position is based on the discussion in Yearley, 1992: 149–81). Dependency on investment from the developed world, for instance, often leads developing countries to relax environmental standards in order to attract multinational companies, which can then avoid the more rigorous standards applied in their home countries (in addition to taking advantage of cheap labour and the availability of raw materials). Likewise, some developing countries, particularly in West Africa, are willing to accept the income generated by the dumping of hazardous wastes which no developed country will accept. The consequence of this drive for income is often excessive pollution and sometimes, as in the disaster at the Union Carbide pesticide plant in Bhopal, an immediate and devastating impact on human life and health.

Agricultural development in the developing world, too, is designed to meet the demands of the developed world. As a consequence, the needs of local people are ignored, many not being able to afford to buy the food produced, and the imported intensive methods employed cause environmental damage. Thus, forests are cleared for cattle grazing and environmental regulations are relaxed so that many products banned in the developed world are used. An additional factor is the loans which developing countries have incurred to industrialise. In the past, it was rare for environmental conditions to be attached to loans and too often the capital was used to invest in large projects, such as dams, which had dire environmental consequences. The environmental consequences of dependency worsened with the onset of the debt crisis in the 1980s. As world trade slumped, developing countries found it increasingly difficult to repay the interest on the massive debts they had been encouraged to take out by the banks. As a consequence, as they made desperate attempts to find the resources to meet their financial commitments, concern for the environment became an even lower priority.

It is not difficult to see, then, why it is commonly argued that principles of justice dictate that developed countries have a moral responsibility towards developing countries in lessening the impact of environmental problems and acting so as to reduce inequalities between the North and the South. The question of moral responsibility for clearing up environmental problems and the environmental consequences of dependence is one thing. The issue of whether reducing inequality is functional for sustainability is quite another. It seems at least equally probable that encouraging economic development in the South will add to environmental problems. As Carter (2007: 63) points out, many environmental problems are the result of affluence.

Clearly, problems such as climate change, ozone depletion and acid rain have all been the product of economic growth and increased consumption in the developed world. Moreover, those with the most disposable income in the developed world have contributed disproportionately to environmental problems. It seems likely, then, that a more equitable global distribution of wealth will not reduce environmental degradation but just redistribute responsibility for it. The developing world explicitly claims its right to gain the material benefits presently limited to the developed world. Yet, it is difficult to deny, on the grounds of an anthropocentric theory of justice, their right to economic development.

The question of global justice is made more complicated, of course, in as far as we are obliged to consider the interests of future generations. There is a good case, as we saw in Chapter 4, for the claim that we do have moral obligations to future generations. If this is the case, there may well be conflicts between the demands of intra-generational and inter-generational theories of justice. That is, if we decide it is a matter of justice to alleviate poverty in the developing world, and this further damages the environment, then what do we say about future generations – those who will have to cope with the environmental degradation we have caused? Some scholars, such as Beckerman and Pasek (2001), argue that benefiting future generations at the expense of the currently living is unjust, since it damages the interests of the poorest people on the planet. In other words, we may be forced to concede that environmental protection is not, at least some of the time, consistent with justice.

Justice and climate change

Political philosophers and moral philosophers have been particularly interested in considering the relationship between intra- and inter-generational theories of justice in the context of climate change (e.g., Page, 2006; Vanderheiden, 2008; Caney, 2008; Gardiner et al., 2010). Climate change has been the major site for conflicts over intra-generational and inter-generational justice. The political reality is that it has proven very difficult for the developed North to persuade the developing South to actively participate in international agreements designed to promote sustainable development. The interrelated issues of environment and development provided the backdrop, for instance, to the United Nations Conference on Environment and Development (the Earth Summit) held in Rio de Janeiro in June 1992. The events of the Earth Summit provide a useful illustration of the conflict between the North and the South on environmental protection. While developed countries were more interested in talking about the pressing environmental issues, developing countries were more interested in development questions and, in particular, action on what they perceived to be the unfair world economic system (Thomas, 1992: 257). As a result, any action to tackle global warming and other environmental problems by the South was to be conditional upon financial assistance from the North. For its part, as the biggest producer of CO_2 emissions – about 23% of the world total from 4% of the world's population – the United States' delegation was adamant that unless developing countries were prepared to, at the very least, indicate a future commitment to cut their own emissions, they would be unable to support a binding treaty because of the loss of economic competitiveness that would occur.

At Rio, legally binding agreements drawn up following extensive prior negotiations were eventually signed on climate change and biological diversity (see Brenton, 1994: 163–95 for the former and 197–206 for the latter). The administration of George Bush senior in the United States refused to sign either convention, however, although this decision was reversed when Clinton succeeded to the Presidency. Moreover, the developing countries were able to sign the Climate Change Convention because the agreement – a reduction in three greenhouse gases (CO_2, CH4 and nitrous oxide) to their 1990 levels by the end of the century – applied only to developed countries. Further, post 2000, cuts were agreed to by developed countries at the follow-up meeting held at Kyoto in December 1997, but again developing countries were not included and the United States was again extremely reluctant to commit to an agreement and only did so at the last minute. To bring the story up to date, despite the fact that developed and developing countries agreed to participate in a post-Kyoto climate change treaty, President Trump subsequently pulled the United States out of the agreement on the grounds that it was detrimental to the American economy.

Negotiations on climate change, then, have continued to be dogged by divisions between the developed and developing worlds. This political reality, of course, explains the prominence of the concept of sustainable development. Originally put on the United Nations (UN) agenda in the 1970s and highlighted by the Brundtland Report, commissioned by the UN Secretary General in 1983 and published in 1987, sustainable development was defined as being about meeting 'the needs of the present without compromising the ability of future generations to meet their own needs' (World Commission on Environment and Development, 1987: 8). It therefore seeks to reconcile the environmental protection demands of inter-generational justice with the development demands of intra-generational justice. By contrast, the Limits to Growth discourse, through its assertion that economic growth must be sacrificed immediately to prevent catastrophic future changes, puts more emphasis on the former (Carter, 2007: 218).

Political theorists have asked three key questions about climate change and justice (Beckman and Page, 2008: 527–28): who are the recipients of climate justice, what are they entitled to receive and who is to bear the financial burden of preventing climate change that will damage the interests of future generations and of adapting to the negative effects of climate change that have already happened?

In terms of the first question, there are intra- and inter-generational dimensions. The environmental justice movement is, in part, concerned with the disproportionate impact that climate change has on the most disadvantaged communities currently living. However, future generations might be regarded as the main recipients of climate change justice. Future generations are particularly threatened by anthropogenic climate change and therefore actions which lead to dangerous levels of greenhouse gas emissions are unjust on those who are yet to be born. Such a conclusion is based on the principle that humans have a right to a healthy environment, and climate change 'jeopardises several fundamental interests', such as interests in subsistence, economic independence and health (Caney, 2008: 538). As with the arguments against inter-generational justice in general, considered earlier in this book (see Chapter 4), a number of objections – based on, for example, reciprocity and lack of knowledge about needs – can be raised against the claim that we are obliged to guarantee the security of future generations.

If we accept that future generations are owed something then what exactly is it that they are owed? Here, an objection, as we saw above, is that the rights of future generations should be discounted because 'the development rights of current generations may override the discounted rights of current and future people not to suffer from dangerous climate change' (Caney, 2008: 539). In other words, less value should be given to the interests of people the further into the future they are born; the so-called 'pure time preference'. If such a principle was accepted then it might be justified morally for the present generation to continue burning fossil fuels with few restraints because their interests take precedence over those of future generations – a view held by, among others, Nordhaus and Boyer (2000). Caney (2008), by contrast, seeks to justify a 'zero discount rate' in which future generations are to receive the same right to a healthy environment as those currently living (see Box 8.4). Those advocating less discounting will advocate stronger measures to combat climate change precisely because of the costs that future generations will have to bear (Gardiner, 2004).

Box 8.4 Objections to Zero-Discounting

Caney raises and rejects a number of objections to his zero-discounting position (2008: 541–51).

- It ignores the preference people have for living for the present and not for the future. However, leaving aside the validity of the empirical claim that people do actually have this preference, it cannot be used to justify the normative assertion that people ought to act in such a way.
- It represents an illegitimate attempt by elites to impose their own normative judgements (that we ought to give equal rights to future generations) on others. However, as Caney (2008: 545–47) points out, this is to ignore the fact that the pure time preference – which argues that the rights of the presently living *ought* to be valued higher than those of future generations – is equally normative.
- It is too demanding on people currently living, since it requires sacrifices that are too demanding (Posner and Weisbach, 2010). However, this is only likely to apply if we maximise the preference satisfaction of future generations. All the zero-discounting position is committed to is protecting basic rights.

The final question about the relationship between justice and climate change is concerned with identifying who it is that should bear the burdens of acting to prevent dangerous levels of climate change and/or mitigating its effects. As with global environmental problems in general, it is common to regard the developed world (its peoples, governments and businesses) as largely responsible for causing the climate change problem. So, according to this 'responsibility' or 'contribution to problem' principle, the developed countries should be responsible for reducing their own greenhouse gas emissions, as well as providing financial support for the developing countries to enable them to grow more sustainably and to adapt to the environmental degradation already caused by climate change. Such responsibility is recognised by the UN and the Kyoto Protocol (the treaty designed to agree cuts in greenhouse gas emissions) which has established the notion of 'common but differentiated responsibilities'. In other words, countries 'should contribute to the costs of managing climate change in proportion to their share of global cumulative greenhouse emissions' (Page, 2008: 557).

There are some problems with the simple allocation of responsibility to developed countries for mitigating, and adapting to, climate change (Page, 2008: 558–59; Pickering and Barry, 2012). In the first place, the practice of regarding states as the entities shouldering the burden for climate change, rather than identifying those specifically responsible within states, is problematic. The effect of so doing might be to penalise poor people within developed states at the expense of the relatively affluent within developing states (Vanderheiden, 2016: 329). In addition, countries in the developed world will soon cease to be the main contributors of greenhouse gas emissions. Even now, the United States has been overtaken by China as the largest single source of greenhouse gases and it is predicted that within 20 years up to three-quarters of emissions will be from developing countries. Moreover, although it is true that the current problems caused by climate change are the result of a build-up of greenhouse gases, mainly from developed countries, over hundreds of years, it can then be asked whether developed countries can be held responsible for effects that could not have been predicted at the time. Until the climate change problem was identified, relatively recently, the burning of fossil fuels could not be regarded as a calculated act of wrong-doing. As a result of this, many political theorists only seek to measure later emissions, Caney (2008), for instance, only measuring post-1990 emissions.

Another problem is that the 'contribution to problem' principle does not take into account the natural advantages and disadvantages involved in variations in national emissions. It would seem unjust, for instance, to penalise a country for high emissions where temperatures are more extreme than elsewhere, therefore requiring more energy for the heating or cooling of buildings (Vanderheiden, 2016: 328). Likewise, if we imagine that climate change is a natural phenomenon, in which case anthropogenic factors are not responsible, then the 'contribution to problem' would suggest that we should leave those affected by it to their fate, since no human agency can be held responsible for it. Our moral intuition, Page suggests, would be that such an outcome is unjust. Moreover, climate change is the result of economic activity over hundreds of years. Applying the 'contribution to problem' approach would therefore target those who are now dead. Why, it could be asked, should their descendants have to pay the price?

An alternative approach for attaching responsibility is ability to pay (Page, 2008: 561). Thus, since developed countries are richer, they should contribute more to resolving climate change. It could be questioned, though, whether it is just to penalise a relatively rich country if its responsibility for climate change is negligible. In practice, of course, rich countries do have responsibility for climate change. Indeed, their wealth has been built upon economic growth driven by the burning of fossil fuels. In abstract terms, though, responsibility would seem to be an important factor. Page's answer (2008: 562) is to suggest a hybrid approach – 'beneficiary pays' – which gives weight to both responsibility and ability to pay. Thus, 'the countries benefiting the most from greenhouse emitting activities in the past bear the greatest responsibility of climate justice'. Based on this ethical theory, attempts have been made to quantify the distribution of financial burdens, a classic example being the Greenhouse Development Rights approach (see http://gdrights.org/about/).

Conclusion

This chapter has considered the extent to which protecting the environment might be regarded as a just thing to do. Given that it is the poorest in the world who suffer most from environmental degradation, there is a case for distributing resources to enable them to relieve this suffering. However theoretically justified a cosmopolitan theory of justice might be, though, the political reality is that in a world of powerful sovereign states it will be difficult to distribute environmental justice to those who ought to receive it (Kamminga, 2008). In short, the rich developed countries have been extremely reluctant to take responsibility for global environmental problems. Thus, an effective climate treaty has stalled on this very question, with developing countries reluctant to act to control greenhouse gas emissions unless developed countries redistribute resources to enable them to cope.

9 THE ENVIRONMENT AND POLITICAL IDEOLOGIES

This chapter – on the relationship between the environment and political ideologies – has been left until now because many of the preceding chapters have, sometimes explicitly and sometimes implicitly, drawn from a variety of ideological traditions. This is not surprising given that all of the major ideological traditions consist of a combination of concepts. It is sometimes said that ideologies are less rigorous and sophisticated than 'proper' political philosophy. In reality, as Vincent (1995: 17) points out: 'ideological themes can be found on a continuum from the most banal jumbled rhetoric up to the most astute theorizing'. This chapter is concerned with codifying the links, many of which have already been identified, between environmental protection and traditional ideologies. It will be asked which conventional ideological tradition – whether liberalism, conservatism, Marxism, socialism, feminism, anarchism or even authoritarianism/fascism – is most compatible with environmental protection or, alternatively, whether environmentalism (or, to be more precise, ecologism) is best seen as a distinct ideology.

Preliminaries

Two preliminary points need to be made. Firstly, at the risk of stating the obvious, whether or not a particular ideological tradition is compatible with environmentalism depends to a large extent on what is meant by the ideology in question. A glance at the literature reveals that virtually all ideological traditions claim to be able to incorporate a concern for the environment. Thus, there are strands of ecosocialism and ecofeminism, and so on. This reflects the fact that ideologies, as a collection of concepts, can be very broad. To be more rigorous, an attempt can be made to identify the core characteristic of a particular ideology.

This is difficult, if not impossible, however, because all ideologies have different strands or schools, and sometimes there is considerable overlap between one ideology and another, such as when we talk about liberal versions of feminism or social democratic versions of socialism. Ideologies are, then, in the words of Festenstein and Kenny (2005: 4), 'internally pluralistic, contested, complex, and overlapping'.

If we focus on particular strands of an ideology, then, is it possible to claim that ideologies as diverse as conservatism, liberalism, Marxism, socialism, feminism and anarchism can, at least to some extent, be made compatible with the protection of the environment? This, in turn, would suggest that there is less resonance in the claim that environmentalism represents a distinct position, not reducible to the political prescriptions of other ideologies. If we adopt the approach to ideologies suggested by Freeden (1996), however, this pluralistic model does not seem quite so compelling. In a major study of the concept of political ideology, Freeden recommends identifying the morphology of an ideology. By this he means it is possible to distinguish between concepts at the core of an ideology from those which are further away from the centre and those which are at the periphery. If we adopt this approach, it might be possible to challenge the environmental claims of a particular ideology.

The second preliminary point is the equally obvious response that the compatibility of environmentalism with traditional ideologies is also dependent upon what is meant by environmentalism. Here, it is important to distinguish between environmentalism and what Dobson labels ecologism (Dobson, 2007) (see Box 9.1). We can ask at least two questions about the relationship between ecologism and other ideologies. In the first place, we can ask the obvious question: how far are traditional ideologies, such as liberalism and socialism, consistent with ecologism? Secondly, and more usefully, we can ask: to what extent can advocates of ecologism learn from traditional ideologies in order to further the goal of environmental protection?

Box 9.1 Environmentalism and Ecology

Environmentalism is a reformist position that sees the protection of the environment as essentially just another policy issue which the political system has to cope with. Although environmental issues may cause peculiar problems which require some institutional tinkering, no fundamental reordering of the political or value system is required and no particular set of political arrangements is

Continued

defended. Seen in such a way, environmentalism is suitable for single-issue pressure groups and for absorption by the mainstream political parties. Its claims to be a distinctive ideology are therefore weak.

By contrast, what Dobson refers to as ecologism is a much more self-contained, and coherent, set of principles which cannot merely be tacked on to an existing set of policies or incorporated within an established system of institutions and values. In its purest form, as we have seen throughout this book, the ecological position advocates a limit to economic growth, rejects an anthropocentric view of the world, and advocates radical social and political change.

In terms of this second question, the issue of power and agency are prominent. In particular, as we saw in earlier chapters, ecologism regards anthropocentrism as the chief cause of environmental degradation. Socialist and feminist accounts, however, have argued that this approach is flawed because it is 'abstracted from a social context' and lacks 'a theory of political power' (Hay, 2002: 27–28) which seeks to explain how exploitative power relations result in environmental degradation. In other words, it is not anthropocentrism that is the major problem but 'the desire of some humans to dominate others' that 'leads to the dominance and degradation of nonhuman nature' (Hay, 2002: 68).

With these two preliminary points, sketched above, in mind, we are now in a position to examine a range of ideological positions. We will start by looking at the two Western ideological giants, liberalism and socialism.

Liberalism

Traditionally, environmental political theorists have regarded liberalism in a negative light. In more recent times, however, this has begun to change. For Meyer (2011: 356–62), this emerging engagement has taken different forms. Firstly, there are those liberal political theorists who have environmental concerns (Wissenburg, 1998, 2006; Hailwood, 2004). Others, however, have engaged with liberalism because of its high status within academic political theory, and/or because it is regarded as the dominant discourse within liberal democratic societies (Dagger, 2006).

The first dimension worth considering in an account of the relationship between liberalism and ecologism is the question of inclusion, or who is to count politically. Here, it can be argued that the inclusion of non-humans within the ethical and political community might be regarded as taking the liberal language of individual rights and obligations to its logical conclusion (Martell, 1994: 141; Eckersley, 1995, 1996). Certainly, as we have seen, contemporary advocates of animal rights and liberation do invoke the liberal tradition of rights and utility to justify their position (see Chapter 5). It is true that both J.S. Mill and Jeremy Bentham recognised that non-human animals, as sentient beings, were entitled to moral consideration, although neither was prepared to say (as consistency, at least for Bentham, demanded they ought) that non-human animals should have a moral status much closer to that of humans than orthodox thinking allowed.

However, as we saw in Chapter 8, liberal social contract theory finds it difficult to justify moral obligations being owed to future generations or non-human animals. In addition, the ecocentric core of Green thinking puts it at odds with the failure of liberalism to countenance going beyond sentience as the grounding for moral standing (and most liberal thought is unashamedly anthropocentric). Moreover, the holistic character of ecological thought clashes with liberal individualism and rights discourse. At the very least, as Hayward (1995: 145) argues, moral arguments might be better couched in terms of responsibilities rather than rights since 'the consistent pursuit of the full range of human rights, for all human beings, would be ecologically unsustainable'.

We can respond in two ways to the difficulty of reconciling ecologism with the liberal emphasis on anthropocentrism and individualism. In terms of the former, eco-minded liberals would point to the possibility of an enlightened anthropocentrism (see Chapter 4). In response to the latter, as we also pointed out in Chapter 4, there are different varieties of holism. So-called holistic value theory – which accords moral value to entities such as ecosystems or species and suggests that their protection is more important than the protection of the individuals within them – is clearly at odds with liberal individualism. By contrast, a holism which suggests – quite sensibly – that the protection of ecosystems or species can serve the function of protecting the individuals within them is more compatible with the liberal focus on the individual.

The traditional hostility to liberalism from advocates of ecologism stems from what is perceived to be the clash between the liberal emphasis on liberty and the demands of environmentalism. As Dobson (2007:

149–50) points out, ecology is predicated on the view that environmental sustainability demands limits to human activities, and this 'amounts to the potential curtailment of certain taken-for-granted freedoms, particularly in the realm of production, consumption and mobility'. That is, 'concern with the environment immediately raises the possibility of interfering with private choices' (Smith, 1998: 51) and therefore challenges the liberty which liberals so prize. Thus, the liberal insistence on state neutrality, that the state should not intervene to favour one particular conception of the good over another, is at odds with ecology's insistence upon a particular conception of the good – one involving a particular moral relationship between humans and nature requiring a focus on protecting the natural environment (Bell, 2002).

One liberal response to this ecological critique is to meet it head-on by emphasising the environmental utility of the free market. There is certainly a debate about the compatibility of the free market and environmentalism (Eckersley, 1993; Vincent, 1993: 267; Wissenburg, 1993). The advocacy of the market has often led liberalism to be regarded as the 'evil genius behind the ecological crisis' (Wissenburg, 2006: 20). Having said that, market solutions to environmental problems have been suggested (Anderson and Leal, 1991; Stavins, 2000). Another argument is that liberal neutrality requires the state to ensure that individuals have the opportunity to further their own conceptions of the good. In order to fulfil this role it is necessary, at the very least, for the state to guarantee the protection of the physical environment: clean water, sustainable energy, preventing damaging climate change, and so forth (Norton, 1999: 119).

Another response that liberals can make is to reject state neutrality. Thus, a revised social liberalism, which from the late 19th century challenged the classical laissez faire model, is much more compatible with environmental protection. This revised version of liberalism justifies considerable state regulation on the grounds that it helps to maximise liberty. From this perspective, then, liberals are not necessarily committed to completely unregulated markets – and the position that owners are entitled to do anything they want with their property – and state intervention on environmental grounds becomes legitimate (Barry, 2001: 66; Stephens, 2001).

One final point to note here is that, even if there is a case for a market approach, or greater state regulation, on liberal grounds, it is a reformist strategy designed to alleviate the worst excesses of economic growth rather than to challenge this growth head-on. As such, it is not consistent with the limits to growth core of ecologism. Despite contemporary attempts to rescue it, it should not be forgotten that liberalism was originally inextricably linked with a productivist ethos. It reached its ascendancy in the 19th century, an era when economic growth, technological optimism and a belief in the ability of humans to master the environment were guiding principles (Gamble, 1981, Chapter 1). One exception was J.S. Mill, who put the case for ecological diversity and a stationary state economy, but he remains very much an isolated figure in the liberal tradition (Eckersley, 1992: 23).

Marxism and socialism

Ideological links between ecology and socialism, widely construed, are commonplace (see, for example, Weston, 1986; Mellor, 2006). This is partly because there have been political alliances between Greens and socialists, particularly in Germany, and partly because Marxism and socialism offer the kind of state or community intervention and regulation which is thought necessary by most environmentalists. Of course, socialism is a very broad ideological church and the answer to the compatibility question depends in part on what one means by the concept (and, indeed, which type of environmentalism one adopts).

In particular, it is necessary, initially, to distinguish between the democratic socialism that emerged in liberal democracies in the 20th century, on the one hand, and the Marxist brand of socialism. The social democratic model shares with all brands of environmentalism an ideological commitment to greater state intervention in the working of the capitalist market. Moreover, it is consistent, too, with the claims of the environmental justice movement which, as we saw in Chapter 8, seeks to show that environmental degradation tends to affect the poorest sections of society most severely. However, democratic socialism focuses on quality of life issues in urban areas and is therefore not particularly concerned with the natural environment in general. Moreover, its goal is greater equality to be achieved through a productivist agenda. Here, the close association between democratic socialist parties and trade unions is likely to be problematic from an environmental perspective since some of the latter tend to be 'organised to protect the interests of labour engaged in environmentally destructive enterprises' (Hay, 2002: 256).

Assessing the compatibility between the Marxist brand of socialism and environmental protection is rather more complex. In the past, Marxists were hostile to environmentalism, seeing it as a middle-class fad, and a means by which the ruling class could divert the attention of the working class away from the importance of the class struggle (Hay, 2002: 259–61). This has gradually changed. Enzensberger (1974) was one of the first to recognise the reality of the global ecological crisis, although he argued that the environmental movement had got its social analysis wrong. Now, there is a considerable body of ecosocialist literature that seeks to reconcile Marxism with ecologism (see, for instance, Barry, 2016; Benton, 1996; Foster, 2000; Gorz, 1982; Kovel 2007; Luke, 2014; Mellos, 1988; Pepper, 1993 and 1993a; Redclift, 1984; Williams, 1986). Ecosocialism of the Marxist variety seeks to argue that the chief cause of environmental degradation is the capitalist system and the class whose interests are served by it. It is the working class, conversely, which is the agency by which capitalism will be overthrown, and, because it is capitalism that is the main cause of the ecological crisis, the working class is also the chief ally of the ecological movement.

What should we make of this claim? Well, much of the focus has been on demonstrating the ecological culpability of the capitalist system of production. In particular, the competitive character of the capitalist system, the need to continually persuade us that we want new goods, and the resulting wastefulness and inefficiency, provide the real reason for environmental degradation (Mellor, 2006: 41–42). As Luke (2014: 206) asserts: 'Marx shows how the dynamics of profit seeking under capitalism violate and degrade both human beings and natural ecologies in the pursuit of monetary gain and material excess.' Kovel (2007) links this ever-increasing search for profits with the phenomenon of globalisation whereby, in a version of Lenin's theory of imperialism, capitalist firms have to sell to new markets thereby spreading ecological degradation further afield. Conversely, under a system of common ownership, production can be organised more effectively and science and technology can be brought to bear in order to provide solutions to environmental problems.

Against this, it has been claimed that a socialist system, in which the means of production are owned in common and the fruits of the productive process are distributed more equally, has no particular ecological benefits. Here, it can be noted that industrialism in China and the former communist regimes of Eastern Europe caused extensive environmental problems (see Gray, 1993: 125–33; Smith, 1998: 73–75). As Scruton (2006: 7) notes, regimes led by the those who claim to be revolutionary socialists have caused untold ecological damage 'as in the forced collectivisation, frenzied industrialization and gargantuan plans to shift populations, rivers and whole landscapes'. Of course, socialists can retort by saying these regimes were not really socialist at all or that environmentally damaging industrialism was necessary because of the threat posed by hostile capitalist enemies (Pepper, 1986: 172–73; Yearley, 1992: 105). What we can say is that, in these countries, the means of production were state-controlled and that state-directed industrialism was not environmentally benign.

The experience of so-called communist regimes is part of the reason why some advocates of ecology have denied that there is an ideological link between their doctrine and socialism, pointing out that the real distinction on ecological grounds is not between capitalism and socialism but between capitalist or socialist industrialism on the one hand and environmental protection on the other. According to this view, it is the 'super-ideology' of industrialism which causes environmental degradation and not a particular form of ownership. As Porritt (1984: 43) points out: 'The politics of the Industrial Age, left, right and centre, is like a three-lane motorway, with different vehicles in different lanes, but all heading in the same direction.' As some members of the British Greens put it, socialism represents 'fair shares in extinction' (quoted in Dobson, 2007: 166).

It is certainly true that some features of Marxist orthodoxy would seem to be at odds with environmental protection. It is, as Eckersley (1992: 24) points out, 'ultimately wedded to the same expansionary ethos and anthropocentric framework as liberalism'. It is anthropocentric because the standard interpretation of Marx suggests that he, just as Locke and Hobbes had done before him, regarded the non-human world exclusively in terms of its use for humans, holding that it only acquired value when being worked on by humans (*ibid*: 25). Even a modern eco-socialist such as David Pepper (1993a: 436–42) accepts that socialism is not ecocentric in orientation and, indeed, he celebrates the fact since ecocentrism, he argues, is often translated into, at worst, anti-humanism and at best a disregard for the human condition – the Gaia hypothesis of Lovelock (2006) being a prime example of the latter.

Of course, if we reject ecocentrism as a fundamental feature of environmental discourse then the Green credentials of Marxism and socialism might be rescued. This is particularly the case if we accept the reinterpretation of the writings of Marx and Engels conducted by scholars such as Parsons (1978) and Benton (1993). Particular attention here is devoted to Marx's early work in which human interaction with the natural environment is regarded as important for spiritual and aesthetic development, as much as it is for meeting material needs (Martell, 1994:

149). Luke (2014), similarly, argues that Marx regarded the protection of the natural world as essential for human beings. The problem here, though, is the difficulty of weighing the importance of this apparent version of an enlightened anthropocentrism in these earlier writings, as compared to what might be regarded as Marx and Engels' more exploitative anthropocentrism.

Leaving aside the debate about the status of anthropocentrism within Marxism, it is less easy to argue away the productivist foundation of socialism and Marxism. The standard interpretation of Marx is that it is not the forces of production (the raw materials, technology and labour power that constitutes the productive process) but the capitalist relations of production which provided the obstacle both to social and political change, and to the unleashing of the full productive potential of the economy. Indeed, the technological developments forthcoming under the capitalist social epoch were the pre-condition for a future socialist society where scarcity would finally be jettisoned (Eckersley, 1992: 77–82). Clearly, such an economic model is seriously at odds with the ecological emphasis on limits to growth and planetary boundaries.

Of course, Marxism is a living ideology and some contemporary Marxists have sought to adapt the doctrine in order to take account of modern concern for the environment. Clearly, there is a great deal in the Marxist, and more general non-Marxian socialist tradition, which ecologists can agree with and perhaps even learn from, and there is a growing body of literature which seeks to draw out these features. A number of themes are worth mentioning here. First is the socialist emphasis on equality. Clearly, redistribution that is driven by economic growth is, from an ecological perspective, unsustainable. However, as Barry (2016: 313) points out redistribution, and therefore greater equality, is more likely to follow from abandoning economic growth. This is because the so-called 'trickle down' process will not then be available as a method of alleviating poverty. Secondly, there is considerable overlap between the vision of small-scale participatory democracies found in the writings of socialists such as Charles Fourier, William Morris and G.D.H. Cole (Hayward, 1995: 177–78) and the preferred social and political arrangements in much ecological thought (Ryle, 1988). Finally, the socialist analysis of political power provides environmentalists with a useful account of how vested interests often stand in the way of environmental improvements, and how environmental protection may be in the interests of the working class, who can be mobilised to promote this objective (see Chapter 10).

Conservatism

Conservatism is rarely seen as an ideology with any affinity to environmental protection. Scruton (2006: 7–8) puts this down to the fact that conservatism is associated with free enterprise and unconstrained markets with humans given the freedom to, if they choose, exploit natural resources. In Britain, this is one of the ideological strands within the Conservative Party, but it is at odds with the defining features of conservatism as an ideology. Indeed, Margaret Thatcher, the chief cheerleader of the free market strand of conservatism, was often described as a liberal.

In reality, there are surprising similarities between conservative thought enshrined, in particular, in the work of the 18th-century writer and politician Edmund Burke, and themes that regularly appear in Green writing (Scruton, 2006; Smith, 1998: 57–61; Wilson, 2016). Greens, like conservatives, are sceptical of the Enlightenment tradition's emphasis on the human capacity to comprehend a complex, interconnected world. Greens, like conservatives, therefore, are persuaded to leave things as they are. Preservation and conservation are, then, terms equally applicable to the conservative and the Green thinker. Present generations are no more than trustees for those still to come.

For conservatives, as for Greens, the appropriate image of a social reformer is a gardener, seeking to prune and trim, rather than an engineer seeking to restructure society according to a rationally thought out plan. These links between Green and conservative ideas enable John Gray (1993: 176), from a conservative perspective, to applaud Green theory, which he argues is 'an invaluable corrective to the Whiggish, anthropocentric, technological optimism by which all the modern political religions are animated and which has, in the form of neo-liberalism, even infected most of what passes today as conservatism'.

There is, on the other hand, much in the conservative tradition which is inconsistent with Green thinking (Dobson, 2007: 158–65). The image of social reformer as 'gardener', for instance, sits rather uneasily with the radical Green critique of 'nature as garden' with its anthropocentric overtones, and the dark Green celebration of wilderness which implies a 'hands-off' approach. Furthermore, conservatism does not have anything to say about the limits to growth, at least beyond a general nostalgia for a pre-industrial past, nor does it seek to extend moral standing beyond humans. Moreover, the radical Green emphasis on radical social and economic change, and the interminable plans of a new society which appear in Green literature, also sit uneasily with the cautious approach of conservatism. Thus, conservatism emphasises the importance of the past as opposed to offering an account of future change. Finally, the modern Green emphasis on egalitarianism and participatory democracy is inconsistent with the importance conservatives attach to hierarchy and authority.

Fascism/Authoritarianism

In the past, the extreme right has shown an interest in environmentalism. Anna Bramwell (1989) has documented the early attention paid to ecology by the Far Right and, indeed, the interest in the far right by ecologists. Elements of the National Socialist party in Germany, for example, displayed some concern with ecological issues from the 1920s and, having gained power, the Nazis were environmentally active, setting up nature reserves and experimenting with alternative forms of energy, among other things. There are ideological similarities between ecologism and fascism too. As Hay (1988: 23), points out: 'Use of biological metaphors, stress on the organic community and the individual's need to merge with it, elevation of ritual, intuition and the mystical, and distrust of the rational' are all elements that can be found in both Green writing and fascist ideology.

However, it is wrong to take the association too far. As Andrew Vincent (1993: 266) comments: 'Because national socialists used socialist methods or favoured ancient German traditions does not mean that either socialism or conservatism are eternally besmirched. The same qualification holds for ecology.' When critics label Green thinking as fascist they are more often than not referring to the authoritarian strain that does exist in some Green writing. This authoritarian strand, particularly prevalent in the 1970s, emphasises the need for drastic coercive solutions – particularly in the areas of reducing consumption and cutting population size – thought necessary to deal with the severity of the environmental crisis (Hardin, 1968; Heilbroner, 1974; Ophuls, 1973). These accounts tend to be based on a Hobbesian model of Man as ultimately selfish and individualistic, requiring an authoritarian state (the Leviathan) to keep order (see Box 9.2).

Box 9.2 Hardin and the Tragedy of the Commons

One well-known exponent of an authoritarian solution to the environmental crisis is Garrett Hardin, whose articles 'The Tragedy of the Commons' and 'Living on a Lifeboat' (1968, 1977) are among the

Continued

best-known environmental writings. His thesis is that, left to their own devices, people will always despoil the commons through greed and naiveté and therefore access to it should be limited by a strong state (or by the institution of private property ownership) to those who are best able to look after it. 'Freedom in a commons', therefore, 'brings ruin to all' (Hardin, 1968: 1244). In the so-called lifeboat ethic parable, he argues, in a Malthusian idiom, that the West (in the lifeboat) should not give a helping hand to the Third World (drowning in the water) since to do so would encourage population growth and greater demands upon natural resources, and so ultimately lead to everyone drowning. Hardin and other like-minded individuals, most notably Paul Ehrlich (1972), formed an organisation called the Environmental Fund in the 1970s which, among other things, campaigned against food aid to the developing world.

Two basic criticisms of Hardin's eco-authoritarianism can be made. In the first place, it might be suggested that his 'Tragedy of the Commons' describes not a commons regime but an open-access regime with no rules or codes of conduct designed to prevent unsustainable exploitation. Further, while most environmentalists would agree with Hardin that environmental damage does come about in the ways he suggests, most would not accept his prescriptions (see Hay, 1988: 23–24; Hayward, 1998: 148–51). This is particularly the case with his extremely illiberal views on the Third World. Indeed, his 'lifeboat ethic' has been criticised for its unjustified pro-developed world stance. As Paehlke (1989: 65–66) puts it:

'Why does Hardin not suggest a more sensible scheme; one advocating the eviction of North Americans from the lifeboat? It would cost far fewer persons and would achieve for the species considerably more time to bring about stabilisation than would avoidable starvation in the so-called Third World'.

Initially, the authoritarian strand was a particular product of the doom-laden 1970s where survival was the major concern; hence the title of Ophuls' work 'Leviathan or Oblivion' (1973). In light of the dominance of liberal democratic regimes, in the West at least, prescribing an authoritarian state might be regarded as just as far-fetched as the decentralisation endemic in contemporary Green writing and activism. To be fair, though, many exponents of the authoritarian position adopt it with a heavy heart, stressing, with some justification, that coercion now is essential to avoid the much greater amount of coercion that will be necessary later to deal with even greater environmental degradation (Martell, 1994: 146). An additional point is that concern about population increases does remain central to environmentalism, but this concern is not synonymous with authoritarianism as a wide variety of non-coercive measures have been recommended.

Feminism

Above all ideologies, it is feminism, perhaps, that has produced the greatest volume of literature on the ecological crisis as well as spawning a burgeoning movement designed to campaign for greater environmental protection. Ecofeminists are very critical of ecologism's targeting of anthropocentrism as the chief cause of environmental degradation. Instead, the real problem is androcentrism (the dominance of a masculine point of view). As Kheel (1990: 128) points out: 'it is not just humans but men and the masculine worldview that must be dismantled from their privileged place'.

The first strand of ecofeminist thought, in the 1970s and early 1980s, was based on the assertion that the exploitation of women and nature were intimately linked as a result of biological differences (see Box 9.3). This biological essentialism has been widely criticised. The simplistic portrayal of different male and female values is questionable, and even if it were possible to attach labels such as 'tough' and 'aggressive' to males and 'maternal' and 'compassionate' to females, it is by no means clear that these are innate as opposed to being a product of the social division of labour (Eckersley, 1992: 66; see also Evans, 1993). As a result, subsequent ecofeminist literature has tended to rely less on biological factors and more on socio-historical factors. Thus, it is not women's biology that enhances their ability to empathise with the natural world, but their historical experience of domination (Griscom, 1981; King, 1981; Plumwood, 1993).

Box 9.3 Ecofeminism and Biological Essentialism

Early ecofeminists adopted a position that was based on the distinctiveness of female biology. Thus, it is masculine values that significantly contribute to environmental degradation, and it is women

Continued

rather than men who have a greater affinity to environmental protection (see Griffen, 1978; Merchant, 1980; Mellor, 1992). Women can identify with the environment since they have a mind-set adjusted to oppression. In other words, the exploitation of nature by men is equivalent to their exploitation of women – hence the title of Collard's (1988) book *Rape of the Wild* (for a feminist account of animal rights along these lines see Donovan and Adams, 2007; Gaard, 1993; see also Regan, 1991, Chapter 5). Similarly, the nurturing role of women (derived from their reproductive function), provides for a set of values (of caring and compassion) more conducive to an identification with the natural world. These values, it is suggested, are inferior to masculine values in Western patriarchal societies and, as a consequence, there is an emphasis on the domination, control, and ultimate despoilation, of nature. According to this argument, then, there is an identity of interests between the ending of male exploitation and the ending of human exploitation of the natural environment.

This revised position is still essentialist, however, merely transferring it from biology to culture (Hay, 2002: 88). As a result, ecofeminism is open to the criticism that it is inaccurate to claim that women have a special empathy with non-human animals and nature. That is, it is not clear that patriarchy and anthropocentrism are synonymous since, as Eckersley (1992: 68) points out, many traditional patriarchal societies did not despoil their natural environment. Likewise, women themselves, particularly in urban environments, have contributed, as much as men, to the domination of nature (Cuomo, 1992; Davion, 1994). Much of the essentialist critique focused on the link drawn by many ecofeminists between the oppression of women and non-human animals, thereby requiring feminists to be vegans or vegetarians (George, 1994). A final point here is the further critique that associating women with non-human animals and nature arguably has the effect of perpetuating their subordination (Plumwood, 2006: 54).

Ecofeminism has always been on the periphery of feminist thought, but, tarred with the essentialism brush, it went into a steep decline, condemned by non-feminists, mainstream feminists and even by some former ecofeminists (Gaard, 2011). Evans (1993: 187) spoke for many when she wrote that given that 'women have no especial, innate tendency towards, or interest in, ecological concerns … it makes sense … to speak …

not of ecofeminists but of ecologists who are feminists too'. Others have insisted on the validity of the links between patriarchy, women and ecological degradation, but have sought to decouple the analysis from essentialism. For some, this has involved dispensing with the ecofeminist label in favour of alternatives such as 'ecological feminism' (Warren, 1994) and 'feminist environmentalism' (Seager, 1993).

Providing a sense of how this more contemporary literature avoids the essentialist critique, and yet provides a distinctive feminist account of environmentalism, however, is difficult. As Hay (2002: 88), rightly I think, points out: 'the question can be asked: if there is no historical or cultural basis upon which a special female empathy with the natural world can be claimed, what is left of ecofeminism?' The most promising non-essentialist ecofeminism centres on the adoption of intersectionality in which 'people can recognise common cause across the boundaries of race, class, gender, sexuality, species, age, ability, nation' (Gaard, 2011: 44). However, the claim that there are linked oppressions, while very convenient, remains under-theorised and lacks sufficient empirical verification (Davis, 2008).

Anarchism

There is also a strong affinity between anarchist thought and much Green political theorising. Anarchism is concerned with justifying stateless, decentralised and self-sufficient communes where people live in harmony with each other (Heywood, 1992: 259–60). Although the ecological imperative was not the major focus of classic anarchist thought, there are obvious parallels between the anarchist vision and the Green account of a post-industrial society. For example, 'on about twenty major points' there is agreement between Peter Kropotkin's *Field, Factories and Workshops* (written in 1898) and the *Blueprint for Survival*, although there is no acknowledgement, and Kropotkin himself was influenced by other anarchists such as Godwin and Proudhon (Pepper, 1986: 188–93).

The best known eco-anarchist is the American Murray Bookchin (1962; 1971), although he is by no means the only author to adopt an anarchist approach to Green political thought (see Schumacher, 1973; Sale, 1984; Bahro, 1986). In a series of books, Bookchin has developed a comprehensive anarchist account (described as 'social ecology') with ecological principles at the forefront (see Eckersley, 1992: 146–60). He suggests that social hierarchy, based on class, race and gender, is the major cause of environmental degradation since domination over non-human nature follows from human domination over other humans. Whether or not one accepts Bookchin's particular brand of eco-anarchism, all of the varieties are based on the fundamental principles of decentralisation and self-sufficiency. As such, they are all subject to the critique, examined in Chapter 6, which questions the ability of such small communities to deal with the international and global nature of environmental problems.

Conclusion: A distinct ideology?

Our relatively brief exploration of the ideology of ecologism in this chapter reveals that it draws its ideas from a wide variety of political traditions. This is not surprising. As Hayward (1995: 199) sensibly suggests, we should not seek 'absolute novelty in conceptualizing society and politics' from ecologism, but rather 'a thoroughgoing critique of dominant conceptions of the political in the light of ecological considerations'.

In order to assess Green thinking more coherently, we can, following Martell (1994, Chapter 5), ask two questions. First, is there anything distinctive about the ideology of ecologism which offers a challenge to mainstream political thinking? Here, the answer must be yes. As Eckersley (1992: 2) points out: 'The questions raised by environmental philosophers have exposed a number of significant blind spots in modern political theory.' Two blind spots are particularly important. First, the emphasis on the limits to growth adds a new dimension which political thought must take account of (and to a certain extent has) when arguing for particular social and political arrangements. Further, Green thinking challenges the anthropocentrism of mainstream Western political thought. As Heywood (1992: 245) points out, traditional ideologies 'merely promise different ways of exploiting nature for the convenience and benefit of humankind'. By contrast, a Green theory of justice requires us to consider the relationship, not only between humans currently living in different parts of the world, but also between those now living and future generations, and, last but not least, between humans and the rest of nature (Dobson, 1998). The adoption of an ecocentric ethic – involving the extension of inherent value beyond the human species to incorporate non-human nature – is clearly a distinct departure in political thought and requires fundamental changes in the way we think about social, legal and political institutions and practices. Even if we are unwilling to accept the validity of an ecocentric ethic, the development of an enlightened anthropocentrism still challenges the Western tradition of political thought which has largely ignored, or been hostile to, questions concerning the moral standing of non-humans.

In terms of our second question – to what extent are there clear ecological reasons or justifications for particular kinds of social and political arrangements? – the distinctive character of ecologism is less clear-cut. Some writers sympathetic to the Green cause argue that ecology is consistent with a wide variety of political structures (Ryle, 1988) while others lean towards the view that particular social and political arrangements are more appropriate than others. Clearly, both cannot be right. The answer might be that both are wrong; that while there are a number of political structures capable of fulfilling ecological objectives, the choice is not completely open (Martell, 1994: 160).

As a final point, it is pertinent to note that some of the recent literature promoting the ecological case for a particular ideology is not, in fact, seeking to show the ways in which that ideology is consistent with existing strands of environmental thinking. Rather, it asks how a particular ideology can be adapted to meet the needs of the ecological challenge (see, for example, Mellor, 2006). Alternatively, it is sometimes suggested that the ideology in question is actually better placed to achieve environmental protection than existing strands of environmental thinking (see, for instance, Plumwood, 2006). In the former case, then, there is a recognition of how traditional ideologies can learn from ecologism, while, in the latter case, it is ecologism which is expected to learn from others. Central here is the recognition, derived from ecosocialism and ecofeminism in particular, that ecologists should not ignore the importance of political power and agency.

10 CONCLUSION: THE IDEA OF A SUSTAINABLE FUTURE AND HOW TO ACHIEVE IT

This book has engaged with the key political ideas pertaining to the human relationship with the natural environment. It has been structured around three key organising themes – interests, values and inclusion. The existence of competing interests and values, and the debate about who should be included as morally considerable and worthy of political inclusion, constitutes the political dimension of environmentalism. What has been revealed is considerable disagreement about what a sustainable society is, how it ought to be achieved, and for whom it is designed. This confirms why a *politics* of the environment is necessary and is likely to continue.

It is clear that the environmental record of governments across the world falls hopelessly short of radical Green objectives, and that, arguably – at least in public – environmentalists should stop expecting governments to achieve them. As Barry (1999: 27) points out: 'The problem with deep ecology is that it brings green politics into irresolvable conflict with settled convictions, giving it a "fundamentalist" complexion which is a hindrance to convincing non-believers to support its political aims.' However, it should also be recognised that until relatively recently, many countries were governed according to a cornucopian ideology, which accepted the validity of unrestrained economic growth and an instrumental attitude towards the exploitation of the natural world, and was naively optimistic about science and technology's ability to come up with solutions to environmental problems (Pearce, 1993: 18–19). It is a mark of how far the world has come that, in rhetoric at least, few governments would now subscribe to this ideology. As John Gray (1993: 155) points out: 'If time travel were possible, a visitor from an earlier period of industrial society … would most likely be astonished by the cleanliness and integrity of our environment, which early industrialism ravaged.'

The pervasiveness of materialism, and anthropocentrism, coupled with a commitment to existing state-based political institutions, means that a reformist environmentalism may well be politically astute. The commitment to economic growth and high standards of living is compatible with ecological modernisation since its central argument, as we saw in Chapter 3, is that conflict between economic and environmental objectives is illusory. Of course, it is not as simple as that. The argument is that, *overall*, the economy will not suffer from the adoption of sustainable policies. Some individual industries, however, will suffer from such policies, at least in the short term.

Likewise, as pointed out in Chapter 4, an enlightened anthropocentrism – where it is insisted that it is in the interest of currently living humans to protect the environment – also offers the prospect of a reconciliation between human chauvinism and sustainability. The inclusion of future generations, non-human animals, and even nature, into democratic theory may well, as was suggested in Chapter 7, be intellectually legitimate. The practical limitations, however, are obvious, and it may well be that encouraging deliberation – and therefore the prospects of greater empathy with the interests of future generations and non-humans– is more politically realistic, albeit still aspirational. Finally, the familiarity of the state as the chief political institution suggests that the best way forward is to attempt to Green state institutions, rather than to promote more imaginative, albeit extremely radical, alternatives.

Green politics, civil society and emancipation

Of course, there are many who still insist that the components of radical Green politics – the limits to growth, the undesirability of materialism, a non-anthropocentric ethic and radical political change – are desirable and indeed essential if a sustainable society is to be created. What radical Greens have to demonstrate, however, is how this change is to come about. Here, it can be argued that focusing on treaty negotiations and international diplomacy between states (as accounts of environmental politics tend to) is to adopt a much too narrow view of what environmental politics is, and ought to be, about. Rather, we ought to be looking at the role of non-state actors, such as transnational environmental activists and community groups of various kinds, and their participation in alternative modes of governance. This is partly an empirical claim, which reflects the fact that such non-state actors are already engaged in alternative forms of environmental governance (Betsill, 2006). It is also, and mainly, a normative argument, that this is what global environmental governance ought to be like, because it is more democratic and participatory and has a higher chance of success.

This leads us away from the focus on states and international negotiations and towards a focus on civil society, and a strategy to create a more environmentally aware citizenry. Environmentalists have not been particularly good at developing such a strategy. There has tended to be an assumption that once the public are informed about the seriousness of environmental problems, they will eventually come round to accepting the need to act (Dobson, 2007: 134–35). Such an assumption is naive, not least because it fails to recognise the power relations central to an understanding of decision-making.

The claim here, then, is that effective environmentalism goes hand in hand with an active citizenry (see Dryzek, 1990). Thus, the move towards a sustainable future requires more than a top-down government response; it also requires a major cultural transformation, and 'a transformation in the beliefs, attitudes and behaviour of individuals' (Carter, 2007: 65). It is in this context that the concept of 'ecological citizenship' has been developed by a number of Green political theorists (Dobson, 2003, 2006; Dobson and Bell, 2006; Hayward, 2006; Light, 2002). For Dobson (2003), ecological citizenship is marked by its cosmopolitan character since, unlike conventional notions of citizenship, its reach is not limited by the nation state. Secondly, it operates in the private as well as the public realm, since what we do as individuals can have a large impact on the environment. Thirdly, it involves the exercise of a number of non-reciprocal obligations on individuals who must be armed by virtues such as justice, care and compassion. This is an alternative to the economic instrument strategy, which accepts that individual behaviour is characterised by self-interest. By contrast, 'the ecological citizen is motivated to live sustainably because it is just to do so, not because she or he is prompted to do so by fiscal incentives or disincentives' (Dobson, 2007: 134).

Green agencies

In this book we have already, of course, considered the role of civil society in environmental politics. In Chapter 7, for instance, we considered the case for regarding democracy, and particularly deliberative notions of democracy, as a vital ingredient of a genuine Green political thought. Here, we can build on this by discussing additional dimensions of civil society's environmental role in terms of lifestyle, communities and class (Dobson, 2007: 119–47).

Lifestyle changes

One seemingly effective way of promoting active environmentally aware citizens is encouraging individuals to adopt sustainable lifestyles. This can be in terms of our behaviour as consumers, as investors, as commuters and in the way we treat the animals we directly encounter. In this way, we can reduce our ecological footprint. Encouraging Green consumerism has been the major component of this strategy. Conserving energy and water through actions such as better insulation and a more discriminate use of the toilet flush is one form of this, as is recycling products bought, a practice which some local authorities have done a great deal to promote.

Most attention, though, has been placed on the Green dimension of regular shopping. Superficially, at least, this strategy – 'shopping for a better world' as it has been called (Porritt and Winner, 1988: 193) – has been very successful. In the late 1980s, there was a huge increase in products claiming to be environmentally friendly – CFC-free aerosols, cruelty-free cosmetics, minimal and biodegradable packaging, 'dolphin-friendly' tuna and free-range eggs – as companies sought to compete with each other to satisfy the new Green demands of the consumer. A book directing consumers to the best Green products became a best seller (Elkington and Hailes, 1988), and retailers specialising in providing such products, most notably the Body Shop and Beauty Without Cruelty, massively increased their market share. Industry has more generally recognised the marketability of environmental concern – witness Shell's 'Better Britain' campaign and Heinz's 'Guardians of the Countryside' programme in the past. Industry too has contributed significantly to the funds of environmental groups. For example, 20% of the World Wide Fund for Nature's income in 1987 came from corporate contributions (Simpson, 1990: 31).

The main advantage of the lifestyle strategy in general, and Green consumerism in particular, is that it focuses on what individuals can do, thereby encouraging participation which, in turn, may lead to an increased consciousness about wider environmental issues. Without wishing to belittle the importance of the consumer strategy, however, it should be noted that there are a number of problems with it. In the first place, the 'Green' claims made by some manufacturers are dubious and it is difficult, at least without legislative intervention, for consumers to interpret the competing claims made. This lack of information provides an incentive for manufacturers to make misleading, or just bogus, statements. An infamous example here is the claim made by many manufacturers of washing up liquids that their products are 'phosphate-free'. True enough, but what they forgot to add was that phosphates had never been added to washing up liquids in the first place! (Yearley, 1992: 98) Similarly, what is a 'cruelty-free' cosmetic can vary from one manufacturer to another. For some companies, it is justifiable to make this claim if they themselves have not done any animal testing (even if the manufacturers of the ingredients have). For others, the label can be applied if products have not been tested for a five-year period. This problem can and has started to be overcome with the introduction of eco-labelling schemes, but this requires governmental action, which may not be forthcoming.

Second, it is doubtful if enough consumers can be persuaded to adopt Green lifestyles, particularly if not doing so is more convenient and/or less expensive. It is a strategy that appeals in particular to those who can afford to engage in substantial consumption. For those who cannot, the options are often limited to buying cheaper but more environmentally damaging products or, in other cases, not consuming at all. In both cases, the individual is locked out of the strategy completely, thus undermining its educative and consciousness-raising purpose. There is also a classic collective action problem here in that the environmental benefits of Green consumerism will be available to all, irrespective of whether any one individual seeks to participate in achieving them. Third, the consumer strategy focuses on the finished product and not the nature of the whole production process. Important environmental questions, such as how the product was manufactured and how it is transported to retailers, are thereby avoided (Yearley, 1992: 101).

While the consumer-oriented strategy is acceptable from a reformist perspective, from a radical Green perspective, of course, it is inadequate because even though it encourages Green growth, it is growth all the same. Radical Greens emphasise the need to reduce consumption, whether it is environmentally friendly consumption or not (Dobson, 2007: 121). Only in this way can a substantial dent be made in our ecological footprint, and only then can the values of a non-materialistic lifestyle be promoted. It is argued, therefore, that environmentalists would do far better to focus their attention on the spiritual dimension of what it is to be Green rather than seeking to promote 'grubby' materialism whether or not it has a Green tint to it (Young, 1993: 105–6).

Communities

Another strategy to increase environmental consciousness is based on the creation of alternative communities working outside of the dominant industrial and materialist paradigm (Dobson, 2007: 123–29). These communities, in the words of the German environmentalist Rudolf Bahro, are 'liberated from the industrial system … liberated from nuclear weapons and from supermarkets. What we are talking about is a new social formation and a different civilisation' (quoted in Dobson, 2007: 123). Examples here would include self-sufficiency farms, some squats and workers' co-operatives; in particular, mention should be made of the Centre for Alternative Technology at Machynlleth in Wales, the Findhorn community in Scotland and the Earth Centre educational project in South Yorkshire.

We can applaud the existence of such communities, but as an agent of social and political change they are surely very limited. Their success in this context will depend upon how far they are able to persuade others to adopt similar lifestyles. Here, they have been a complete failure. This is partly because such communities tend to be 'outsiders' rather than 'oppositional' in that they represent opting out of society as opposed to deliberately setting themselves in a confrontational sense against the norms and values of industrial society (Dobson, 2007: 124–25). Whatever the reason, while laudable by themselves, they do not represent a convincing strategy for achieving a widespread raising of environmental consciousness.

A privileged subject of change

A more promising strategy is the identification of a particular group or class in society who can be mobilised on behalf of environmental objectives. David Pepper (1986: 215–25) describes what he calls education *from* the environment as an approach that introduces values to the debate by asking why we ought to protect the environment. This is something that environmental thinkers and activists have not been shy about. However, it is still inadequate because it assumes that environmental problems represent a common threat against which we should all pull together.

Put starkly, environmental degradation affects different people in different ways. It was emphasised in the introduction to this book, for instance, that climate change has a varying impact on different regions and individuals within the same region. Likewise, the environmental justice movement, encountered in Chapter 8, is built on the premise that environmental degradation hits the poorest hardest. Failure to recognise this more widely means that environmentalists might be accused of being utopian. 'In light of this', as Dobson (2007: 137–38) points out, 'class theory has it that radical greens must abandon their utopian, universalistic strategy, and instead identify and organize a group of people in society whose immediate interests lie in living sustainable lives, with all that implies'. This insight is the basis, as we saw in Chapter 9, of the ecosocialist and ecofeminist critique of ecologism.

Swyngedouw (2009), for one, has argued that the failure to identify a so-called privileged subject of change accounts for environmentalism being depoliticised. Compare this with other social movements – such as the labour movement, feminism, and civil rights – where the subject of oppression as well as the agent for change are easily identifiable. It is important to recognise here that the lack of a privileged subject of change does not render environmentalism non-political. Rather, it prevents Green ideas from germinating because of the absence of a natural constituency. This reflects the fact, previously unjustly neglected, that politics is about who has the *power* to ensure that their interest and values predominate.

It is important, then, to identify and mobilise a collectivity in society that is more likely to lead a cultural transformation in favour of environmental protection. For Pepper, writing from a socialist perspective, it is the class-based power structure which has to be tackled if environmental degradation is to be reversed. The logic of this argument is that the working class are the obvious agents of any social transformation, not least because there is much evidence to suggest that 'the processes of environmental degradation almost always impact most devastatingly on the poorest and least powerful communities, both within countries and globally' (Benton, 1997: 43; see also Huby, 1998 and Monbiot, 1999). By contrast, the middle classes are not affected by environmental degradation to the same degree as the less affluent. The more affluent are able to escape from the grime of the inner cities and, when encamped in more salubrious locations, they can exercise considerable political clout through the party system and, moreover, are articulate enough to influence the planning system. As a result, they can ensure that their immediate environment remains reasonably good.

The converse view is that, precisely because of this political leverage and the desire for a protected environment, the middle class, or at least a section of it, might be an effective Green agency. This claim is reinforced by Inglehart's claim, discussed in Chapter 2, that post-material values are a product of affluence, and further evidence is that the Green Party's support tends to be drawn from the more affluent sections of society. As Porritt (1984: 116) points out:

> the post-industrial revolution is likely to be pioneered by middle-class people ... such people not only have more chance of working out where their own *genuine* self-interest lies, but they also have the flexibility and security to act upon such insights.

On the other hand, the middle classes have, despite Hirsch's identification of the positional economy, benefited from sometimes unsustainable economic growth. Moreover, their adoption of a suburban or rural lifestyle is essentially shallow since it is based upon material convenience and, in so far as the middle classes do engage in environmental campaigns, it is more often than not designed to protect their own privileges against encroachment by others.

If the middle class is not a particularly likely agent for environmental change, does the working class fare any better? Arguably not. For one thing, in Britain at least, the working class has fragmented in recent years and it can no longer be regarded as a homogeneous block. The new, more affluent, sections of the working class, seduced by the materialistic promises of successive governments, are therefore unlikely to be a vehicle for environmental protection. Furthermore, the organised labour movement, in Britain at least, has always been 'labourist' in character, concerned, despite the strain of ethical socialism associated with such luminaries as William Morris and R.H. Tawney, with gaining for itself a larger slice of the capitalist pie rather than seeking to change the acquisitive and materialist nature of the system.

Two other social groupings suggest themselves as Green agents. The first is the women's movement. As we saw in Chapter 9, there has been an attempt to develop a feminist strand of ecology and part of this is concerned with explaining why women might be more likely than men to identify with environmental concerns. To talk about women as a homogeneous group, though, is problematic since, by virtue of their class position, they experience different life chances. Furthermore, as Chapter 2 revealed, the empirical evidence suggests that affluence and occupation, rather than gender, provide important influences upon the adoption of post-material values. On the other hand, support for environmental reforms is not dependent on holding such values and the life position of women (as, for instance, carers of children) may make them more likely than men to support some environmental initiatives, such as those relating to transport policy.

The final possibility is the unemployed, a social grouping which seems to offer great potential as an agent of environmentalism (a view advanced by Gorz, 1985). This is a group which is relatively disengaged from industrial society and which, because of its material limitations, is already living a much more sustainable lifestyle. In particular, the long-term unemployed have little stake in society and their predicament can be explained by the limits to growth in that unemployment is at least partly a product of the increasing price of scarce resources necessitating cuts in labour costs.

There are problems with according to the unemployed a role as the agents of social change. Crucially, there is little evidence that the unemployed perceive their role as agents of such a change. The Green movement could attempt to make them more aware of their marginalised status and how it relates to the environmental crisis but, in the past, it has been notoriously difficult to mobilise the unemployed to campaign against their predicament. This is associated with the point that there is no necessary link between the marginalisation of the unemployed and their rejection of the social and political institutions of unsustainable industrial society. Indeed, if anything, the evidence suggests that those at the bottom of the social scale working in casual jobs or not at all have been, traditionally at least, more likely to be deferential towards the established order (Nordlinger, 1967).

The point, then, is that marginalisation is just as likely, without the introduction of any evidence to the contrary, to lead to a desire for inclusion rather than the development of an oppositional mentality. Moreover, it is doubtful if the number of unemployed, and the poor in general, in the developed world at least, constitute a large enough group to make a significant impact. Finally, although there is evidence that the unemployed have played a significant role in environmental protest, there is a problem of causation. It seems likely that it is not unemployment which leads to the development of an environmental consciousness, but a concern for the environment and a wish to pursue a life untainted by material excesses, that makes unemployment attractive (Taylor, 1999).

Global change

The cultural transformation required to break the hold that certain economic interests have over environmental policy-making has a long way to go. The promotion of radical Green values which challenge the dominant materialism of industrial society obviously has an important role to play in the development of a greater environmental consciousness. Equally, if not more important, is the ecological modernisation claim that environmental protection and economic advancement can co-exist. Pursuing such an approach may not answer all of the questions posed by radical Greens but it is politically realistic and might just be enough to provide a viable future for the human species and the natural environment we depend upon.

Pursuing environmental protection in the developed world is also a prerequisite for persuading the developing world to act. Without the participation of the latter in a global effort, there is little chance of a sustainable future. While the peoples of the rich North may be prepared to accept some economic sacrifice in return for environmental protection, in the developing world, as Brundtland recognised, only a strategy which promotes the long-term economic benefits of sustainability will succeed, and, in the interests of justice, deserves to succeed.

This book has, above all, sought to demonstrate the dichotomy between, on the one hand, the widespread recognition of environmental problems and an environmental movement bristling with both scientific expertise and moral indignation at the present state of affairs and, on the other, the very slow progress that has been made towards changing people's attitudes and persuading public authorities to introduce more sustainable environmental measures. The spark that seemed to ignite widespread concern for the environment in the mid-to-late 1980s has failed to keep the fire alight and yet the problems remain acute. As an important political issue and a subject worthy of study by natural and social scientists, however, the environment is here to stay. Those who despair at the absence of fundamental change should remember the old adage that the perfect should not be allowed to be the enemy of the good.

BIBLIOGRAPHY

Abramovitz, J. (1997), 'Valuing nature's services' in L.R. Brown et al. (eds.) *State of the World 1997* (New York: Norton), 95–114.

Anderson, T. and Leal, D. (1991) *Free Market Environmentalism* (Boulder, CO: Westview Press).

Asafu-Adjaye, J. et al. (2015) *An Ecomodernist Manifesto*, www.ecomodernism.org/

Attfield, R. (2003) *Environmental Ethics* (Cambridge: Polity).

Attfield, R. (2014) *Environmental Ethics* (Cambridge: Polity, second edition).

Bachrach, P. (1969) *The Theory of Democratic Elitism: A Critique* (London: University of London).

Bahro, R. (1986) *Building the Green Movement* (London: Heretic).

Baker, S. (2006) *Sustainable Development* (London: Routledge).

Ball, T. (2006) 'Democracy' in A. Dobson and R. Eckersley (eds.) *Political Theory and the Ecological Challenge* (Cambridge: Cambridge University Press), 131–47.

Barabas, J. (2004) 'How Deliberation Affects Policy Opinions', *American Political Science Review*, 98, 4: 687–701.

Barnett, J. (2001) *The Meaning of Environmental Security: Ecological Politics and Policy in the New Security Era* (London: Zed Books).

Barry, B. (1995) *Justice as Impartiality* (Oxford: Clarendon Press).

Barry, B. (2005) *Why Social Justice Matters* (Cambridge: Polity Press).

Barry, J. (1994) 'The limits of the shallow and the deep: Green politics, philosophy and praxis', *Environmental Politics*, 3, 3: 369–94.

Barry, J. (1999) *Rethinking Green Politics: Nature, Virtue and Progress* (London: Sage).

Barry, J. (2001) 'Greening liberal democracy: practice, theory and political economy' in J. Barry and M. Wissenburg (eds.) *Sustaining Liberal Democracy* (New York: Palgrave), 59–80.

Barry, J. (2016) 'Green Political Economy: Beyond Orthodox Undifferentiated Economic Growth as a Permanent Feature of the Economy' in T. Gabrielson, C. Hall, J. Meyer, and Schlosberg (eds.) *The Oxford Handbook of Environmental Political Theory* (Oxford: Oxford University Press), 304–17.

Barry, J. and Eckersley, R. (eds.) (2005) *The State and the Global Ecological Crisis* (Cambridge, MA: MIT Press).

Baxter, B. (2000) 'Ecological Justice and Justice as Impartiality', *Environmental Politics*, 9, 3: 43–64.

Baxter, B. (2005) *A Theory of Ecological Justice* (London: Routledge).

Beck, U. (1992) *Risk Society: Towards a New Modernity* (London: Sage).

Beckerman, W. (1974) *In Defence of Economic Growth* (London: Cape).

Beckerman, W. (1992) 'Global warming and international action: An economic perspective' in A. Hurrell and B. Kingsbury (eds.) *The International Politics of the Environment* (Oxford: Clarendon Press), 253–89.

Beckerman, W. (1995) *Small is Stupid: Blowing the Whistle on the Greens* (London: Duckworth).

Beckerman, W. and Pasek, J. (2001) *Justice, Posterity and the Environment* (Oxford: Oxford University Press).

Beckman, L. and Page, E. (2008) 'Perspectives on justice, democracy and global climate change', *Environmental Politics*, 17, 4: 527–35.

Beeson, M. (2010) 'The Coming of Environmental Authoritarianism', *Environmental Politics*, 19, 2: 276–94.

Beeson, M. (2016) 'Environmentalism Authoritarianism and China' in T. Gabrielson, C. Hall, J. Meyer and D. Schlosberg (eds.) *The Oxford Handbook of Environmental Political Theory* (Oxford: Oxford University Press), 520–32.

Beitz, C. (1979) *Political Theory and International Relations* (Princeton, NJ: Princeton University Press).

Bell, D. (2002) 'How Can Political Liberals be Environmentalists', *Political Studies*, 50, 4: 703–24.

Benhabib, S. (ed.) (1996) *Democracy and Difference* (Princeton, NJ: Princeton University Press).

Bennie, L. (1998) 'Brent Spar, Atlantic oil and Greenpeace' in F.F. Ridley and G. Jordan (eds.) *Protest Politics: Cause Groups and Campaigns* (Oxford: Oxford University Press), 89–102.

Bentham, J. (1948) *An Introduction to the Principles of Morals and Legislation* (New York: Hafner Press).

Benton, T. (1993) *Natural Relations: Ecology, Animal Rights and Social Justice* (London: Verso).

Benton, T. (ed.) (1996) *The Greening of Marxism* (New York: Guilford).

Benton, T. (1997) 'Beyond left and right? Ecological politics, capitalism and modernity' in M. Jacobs (ed.) *Greening the Millennium: The New Politics of the Environment* (Oxford: Blackwell), 34–46.

Bernstein, S. (2001) *The Compromise of Liberal Environmentalism* (New York: Columbia University Press).

Betsill, M. (2006) 'Transnational actors in international environmental politics' in M. Betsill, K. Hochstetler and D. Stevis (eds.) *Palgrave Advances in International Environmental Politics* (Basingstoke: Palgrave Macmillan), 172–202.

Blair, T. (2005) Tony Blair's conference speech 2005. *The Guardian*, 27 September 2005.

Blanchard, E. (2010) 'Modelling the Future: an overview of the "Limits to Growth" debate', *Centaurus*, 52, 2: 91–116.

Boardman, R. (1981) *International Organisations and the Conservation of Nature* (London: Macmillan).

Bohman, J. (1998) 'Survey Article: The Coming of Age of Deliberative Democracy', *The Journal of Political Philosophy*, 6, 4: 400–25.

Bookchin, M. (1962) *Our Synthetic Environment* (New York: Knopf).

Bookchin, M. (1971) *Post Scarcity Anarchism* (Berkeley: Ramparts).

Bramwell, A. (1989) *Ecology in the Twentieth Century* (New Haven, CT: Yale University Press).

Brenton, T. (1994) *The Greening of Machiavelli: The Evolution of International Environmental Politics* (London: Earthscan).

Bretherton, C. (1996) 'Gender and environmental change: Are women the key to safeguarding the planet?' in J. Vogler and M. Imber (eds.) *The Environment and International Relations* (London: Routledge), 99–119.

Bullard, R. (2000) Dumping in Dixie: Race, Class and Environmental Quality (Boulder, CO: Westview Press, third edition).

Callicott, J.B. (1980) 'Animal liberation: A triangular affair', *Environmental Ethics*, 2, 4: 311–38.

Callicott, J. (1988) 'Animal Liberation and Environmental Ethics: Back Together Again', *Between the Species*, 4, 3: 163–69.

Callicott, J. (1995) 'Animal Liberation: A Triangular Affair' in R. Elliot (ed.) *Environmental Ethics* (Oxford: Oxford University Press), 29–59. Revised version of the article first published in 1980.

Callicott, J. (2013) *Thinking Like a Planet: The Land Ethic and the Earth Ethic* (New York: Oxford University Press).

Campbell, T. (1988) *Justice* (Basingstoke: Macmillan).

Caney, S. (2008) 'Human rights, climate change and discounting', *Environmental Politics*, 17, 4: 536–55.

Caney, S. (2016) 'Cosmopolitanism and the Environment' in T. Gabrielson, C. Hall, J. Meyer and D. Schlosberg (eds.) *The Oxford Handbook of Environmental Political Theory* (Oxford: Oxford University Press), 238–53.

Carmin, J. and Agyeman, J. (eds.) (2011) *Environmental Inequalities Beyond Borders: Local Perspectives on Global Injustices* (Cambridge, MA: MIT Press).

Carrington, D. (2016) 'The Anthropocene Epoch: scientists declare dawn of human-influenced age', *The Guardian*, 29 August.

Carson, R. (1962) *Silent Spring* (New York: Fawcett Crest).

Carter, N. (2007) *The Politics of the Environment: Ideas, Activism, Policy* (Cambridge: Cambridge University Press, second edition).

Cavalieri, P. and Singer, P. (eds.) (1983) *The Great Ape Project: Equality Beyond Humanity* (London: Fourth Estate).

Chambers, S. (2003) 'Deliberative Democratic Theory', *Annual Review of Political Science*, 6: 307–26.

Chasek, P., Downie, D. and Welsh Brown, J. (2010) *Global Environmental Politics* (Philadelphia, PA: Westview Press, fifth edition).

Cherfas, J. (1988) *The Hunting of the Whale* (London: The Bodley Head).

Christoff, P. (2005) 'Out of Chaos, a Shining Star' in J. Barry and R. Eckersley (eds.) *The State and the Global Ecological Crisis* (Cambridge, MA: MIT Press), 25–52.

Cochrane, A. (2009), 'Do Animals Have an Interest in Liberty?', *Political Studies*, 57: 660–79.

Cochrane, A. (2012a) *Animal Rights Without Liberation* (New York: Columbia University Press).

Cochrane, A. (2012b) 'Cosmozoopolis: The case Against Group-Differentiated Animal Rights', *Law, Ethics and Philosophy*. http://leap-journal.com/archives/LEAP1-Alasdair-Cochrane.pdf

Cochrane, A., Garner, R. and O'Sullivan, S. (2016) 'Animal ethics and the political', *Critical Review of International Social and Political Philosophy*, DOI: 10.1080/13698230.2016.1194583: 1–17.

Cohen, C. (1986) 'The Case for the Use of Animals in Biomedical Research', *New England Journal of Medicine*, 315, 14: 865–70.

Cole, H. (1973) *Thinking about the Future: A Critique of the Limits to Growth* (Brighton: Sussex University Press).

Collard, A. (1988) *Rape of the Wild* (London: The Women's Press).

Connelly, J. and Smith, G. (1999) *Politics and the Environment: From Theory to Practice* (London: Routledge).

Connelly, J., Smith, G., Benson, D. and Saunders, C. (2012) *Politics and the Environment: From Theory to Practice* (London: Routledge, third edition).

Cook, I. and Swyngedouw, E. (2012) 'Cities, social cohesion and the environment: Towards a future research agenda', *Urban Studies*, 49, 9: 1959–79.

Cooke, S. (2017) 'Animal Kingdoms: on habitat rights for wild animals', *Environmental Values*, 26, 1: 53–72.

Cotgrove, S. (1982) *Catastrophe or Cornucopia* (Chichester: Wiley).

Cotgrove, S. and Duff, A. (1980) 'Environmentalism, middle class radicalism and politics', *Sociological Review*, 28: 333–51.

Cotgrove, S. and Duff, A. (1981) 'Environmentalism, values, and social change', *British Journal of Sociology*, 32, 1: 92–110.

Cripps, E. (2010) 'Saving the polar bear, saving the world: Can the capabilities approach do justice to humans, animals and ecosystems?', *Res Publica*, 16, 1: 1–22.

Crutzen, P. and Stoermer, F. (2000) 'The Anthropocene', Global *Change Newsletter*, 41: 17–18.

Cuomo, C. (1992) 'Unravelling the Problems in Ecofeminism', *Environmental Ethics*, 14: 351–63.

Dagger, R. (2006) 'Freedom and Rights' in A. Dobson and R. Eckersley (eds.) *Political Theory and the Ecological Challenge* (Cambridge: Cambridge University Press), 200–15.

Dahl, R. (1979) 'Procedural Democracy' in P. Laslett and J. Fishkin (eds.) *Philosophy, Politics and Society* (Oxford: Basil Blackwell, fifth edition).

Dahl, R. (1989) *Democracy and its Critics* (New Haven, CT: Yale University Press).

Dalby, S. (2002) *Environmental Security* (Minneapolis, MN: University of Minnesota Press).

Daly, H. (1974) 'The Economics of the Steady State', *The American Economic Review*, 64, 2: 15–21.

Davidson, S. (2009) 'Ecoanarchism: A Critical Defence', *Journal of Political Ideologies*, 14, 1: 47–67.

Davion, V. (1994) 'Is ecofeminism feminist?' in K. Warren (ed.) *Ecological Feminism* (London: Routledge), 8–28.

Davis, K. (2008) 'Intersectionality as buzzword: A sociology of science perspective on what makes a feminist theory successful', *Feminist Theory*, 9, 1: 67–85.

Demaria, F. et al. (2013) 'What is Degrowth? From an Activist Slogan to a Social Movement', *Environmental Values*, 22, 2: 191–215.

Descartes, R. (1912) 'Discourse' in J. Veitch (ed.) *Rene Descartes: A Discourse on Method* (London: Dent), 90–97.

Diamond, I. (1994) *Fertile Ground: Women, Earth and the Limits of Control* (Boston: Beacon Press).

Dickson, L. and McCulloch, A. (1996) 'Shell, the Brent Spar and Greenpeace: A Doomed Tryst?', *Environmental Politics*, 5, 1: 122–29.

Dobson, A. (ed.) (1991) *The Green Reader* (London: Andre Deutsch).

Dobson, A. (1996) 'Democratising green theory: Preconditions and principles' in B. Doherty and M. de Geus (eds.) *Democracy and Green Political Thought* (London: Routledge), 132–48.

Dobson, A. (1998) *Justice and the Environment: Conceptions of Environmental Sustainability and Theories of Distributive Justice* (Oxford: Oxford University Press).

Dobson, A. (2003) *Citizenship and the Environment* (Oxford: Oxford University Press).

Dobson, A. (2006) 'Ecological Citizenship: A Defence', *Environmental Politics*, 15, 3: 447–51.

Dobson, A. (2007) *Green Political Thought* (London: Routledge, fourth edition).

Dobson, A. (2010) 'Listening: The New Democratic Deficit', *Political Studies*, 60: 843–59.

Dobson, A. (2016) 'Are There Limits to Growth?' in T. Gabrielson, C. Hall, J. Meyer and D. Schlosberg (eds.) *The Oxford Handbook of Environmental Political Theory* (Oxford: Oxford University Press), 289–303.

Dobson, A. and Bell, D. (eds.) (2006) *Environmental Citizenship* (Cambridge, MA: MIT Press), 62–75.

Doherty, B. (2002) *Ideas and Actions in the Green Movement* (London: Routledge).

Dombrowski, D. (1997) *Babies and Beasts: The Argument from Marginal Cases* (Chicago, IL: University of Illinois Press).

Donaldson, S. and Kymlicka, W. (2011) *Zoopolis: A Political Theory of Animal Rights* (New York: Oxford University Press).

Donovan, J. and Adams, C. (eds.) (2007) *The Feminist Care Tradition in Animal Ethics* (New York: Columbia University Press).

Dryzek, J. (1987) *Rational Ecology* (Oxford: Blackwells).

Dryzek, J. (1990) *Discursive Democracy: Politics, Policy and Political Science* (Cambridge: Cambridge University Press).

Dryzek, J. (2000) *Deliberative Democracy and Beyond: Liberals, Critics and Contestations* (Oxford: Oxford University Press).

Dryzek, J. (2010) *Foundations and Frontiers of Deliberative Governance* (Oxford: Oxford University Press).
Dryzek, J. (2013) *The Politics of the Earth: Environmental Discourses* (Oxford: Oxford University Press, third edition).
Dryzek, J., Downes, D., Hunold, C. and Schlosberg, D. (2003) *Green States and Social Movements* (Oxford: Oxford University Press).
Duit, A., Feindt, P. and Meadowcroft, J. (2016) 'Growing Leviathan: the rise of the environmental state?', *Environmental Politics*, 25, 1: 1–23.
Eckersley, R. (1989) 'Green politics and the new class: Selfishness or virtue?' *Political Studies*, 2, 37: 205–23.
Eckersley, R. (1992) *Environmentalism and Political Theory* (London: UCL).
Eckersley, R. (1993) 'Free market environmentalism: friend or foe?', *Environmental Politics*, 2, 1: 1–19.
Eckersley, R. (1995) 'Liberal democracy and the rights of nature: The struggle for inclusion' in F. Matthews (ed.) *Ecology and Democracy*, special edition of *Environmental Politics*, 4, 4: 169–98.
Eckersley, R. (1996) 'Greening liberal democracy: The rights discourse revisited' in B. Doherty and M. de Geus (eds.) *Democracy and Green Political Thought* (London: Routledge), 212–36.
Eckersley, R. (1999) 'The discourse ethic and the problem of representing nature', *Environmental Politics*, 8, 2: 24–49.
Eckersley, R. (2004) *The Green State* (Cambridge, MA: MIT Press).
Eckersley, R. (2010) 'Green theory' in T. Dunne, M. Kurki and S. Smith (eds.) *International Relations Theories: Discipline and Diversity* (Oxford: Oxford University Press), 257–77.
Ehrlich, P. (1972) *The Population Bomb* (London: Pan).
Ekeli, K. (2005) 'Giving a Voice to Posterity – Deliberative Democracy and Representation of Future People', *Journal of Agricultural and Environmental Ethics*, 18: 429–50.
Ekeli, K. (2009) 'Constitutional experiments: Representing future generations through submajority rules', *Journal of Political Philosophy*, 17, 4: 440–61.
Elkington, J. and Hailes, J. (1988) *The Green Consumer Guide* (London: Gollancz).
Elliot, L. (2004) *The Global Politics of the Environment* (Basingstoke: Macmillan, second edition).
Elliot, R. (1995) 'Introduction' in R. Elliot (ed.) *Environmental Ethics* (Oxford: Oxford University Press), 1–20.
Elstub, S. (2008) *Towards a Deliberative and Associational Democracy* (Edinburgh: Edinburgh University Press).
Elstub, S. (2014) 'Mini-publics: Issues and Cases' in S. Elstub and P. McLaverty (eds.) *Deliberative Democracy: Issues and Cases* (Edinburgh: Edinburgh University Press), 166–88.
Enzensberger, H. (1974) 'A critique of political ecology', *New Left Review*, 84: 3–31.

Easterlin, R. (1974) 'Does economic growth improve the human lot?' in P. David and M. Reder (eds.) *Nations and Households in Economic Growth; Essays in Honour of Moses Abramovitz* (London: Academic Press), 89–125.

Estland, D. (1997) 'Beyond Fairness and Deliberation: The Epistemic Dimension of Democratic Authority' in J. Bohman and W. Rehg (eds.) *Deliberative Democracy: Essays on Reason and Politics* (Cambridge, MA: MIT Press), 173–204.

Evans, J. (1993) 'Ecofeminism and the politics of the gendered self' in A. Dobson and P. Lucardie (eds.) *The Politics of Nature: Explorations in Green Political Theory* (London: Routledge), 177–89.

Evans, P., D. Rueschemeyer and T. Skocpol (eds.) (1985) *Bringing the State Back In* (Cambridge: Cambridge University Press).

Farrelly, C. (2005) 'Making Deliberative Democracy a More Practical Political Ideal', *European Journal of Political Theory*, 4, 2: 200–08.

Farrelly, C. (2007) 'Justice in Ideal Theory: A Refutation', *Political Studies*, 55: 844–64.

Festenstein, M. and Kenny, M. (2005) *Political Ideologies* (Oxford: Oxford University Press).

Fishkin, J. (2009) *When the People Speak: Deliberative Democracy and Public Consultation* (Oxford: Oxford University Press).

Foster, J. (2000) *Marx's Ecology: Materialism and Nature* (New York: Monthly Review Press).

Fox, W. (1984) 'Deep ecology: A new philosophy of our times', *The Ecologist*, 14, 5: 199–200.

Fox, W. (1995) *Toward a Transpersonal Ecology: Developing New for Environmentalism* (Totnes: Resurgence).

Freeden, M. (1996) *Ideologies and Political Theory* (Oxford: Oxford University Press).

Freeman, S. (2000) 'Deliberative Democracy: A Sympathetic Comment', *Philosophy and Public Affairs*, 29, 4: 371–418.

Frey, R.K. (1983) *Rights, Killing and Suffering* (Oxford: Clarendon Press).

Frey, R. (1987) 'Autonomy and the Value of Animal Life', *Monist*, 70: 50–63.

Friberg-Fernos, H. and Karlsson, S. (2014) 'The Consensus Paradox: Does Deliberative Agreement Impede Rational Discourse?', *Political Studies*, 62: 99–116.

Fung, A. (2007) 'Minipublics: Deliberative Designs and their Consequences' in S. Rosenberg (ed.) *Deliberation, Participation and Democracy: Can the People Govern* (Basingstoke: Palgrave Macmillan), 159–83.

Gaard, G. (1993) *Ecofeminism: Women, Animals, Nature* (Philadelphia, PA: Temple University Press).

Gaard, G. (2011) 'Ecofeminism Revisited: Rejecting Essentialism and Re-Placing Species in a Material Feminist Environmentalism', *Feminist Formations*, 23, 2: 26–53.

Gabrielson, T., Hall, C., Meyer, J. and Schlosberg, D. (2016), 'Introducing Environmental Political Theory' in T. Gabrielson, C. Hall, J. Meyer and D. Schlosberg (eds.) *The Oxford Handbook of Environmental Political Theory* (Oxford: Oxford University Press), 3–14.

Gamble, A. (1981) *An Introduction to Modern Social and Political Thought* (London: Macmillan).

Gardiner, S. (2004) 'Ethics and Global Climate Change', *Ethics*, 114: 555–600.

Gardiner, S., Caney, S., Jamieson, D. and Shue, H. (eds.) (2010) *Climate Change: Essential Readings* (Oxford: Oxford University Press).

Garner, R. (2004) *Animals, Politics and* Morality (Manchester: Manchester University Press, second edition).

Garner, R. (2011) 'Animal Welfare, Ethics and the Work of the International Whaling Commission', *Journal of Global Ethics*, 7, 3: 279–90.

Garner, R. (2012) 'Rawls, Animals and Justice: New Literature, Same Response', *Res Publica*, 18, 2: 159–72.

Garner, R. (2013) *A Theory of Justice for Animals: Animal Rights in a Non-ideal World* (New York: Oxford University Press).

Garner, R. (2016) 'Animal rights and the deliberative turn in democratic theory', *European Journal of Political Theory*: DOI: 10.1177/1474885116630937

Garner, R. (2017) 'Animals and Democratic Theory: Beyond an Anthropocentric Approach', *Contemporary Political Theory*: 16, 4: 459–77.

Garner, R. and O'Sullivan, S. (eds.) (2016) *The Political Turn in Animal Ethics* (London: Rowman and Littlefield International).

Garrard, G, Handwerk, G. and Wilke, S. (2014) 'Imagining Anew: Challenges of Representing the Anthropocene', *Environmental Humanities*, 5, 1: 149–53.

Gaus, G. (1997) 'Reason, Justification, and Consensus: Why Democracy Can't Have it All' in J. Bohman and W. Rehg (eds.) *Deliberative Democracy: Essays on Reason and Politics* (Cambridge, MA: MIT Press), 205–42.

George, K. (1994) 'Should Feminists be Vegetarians?', *Signs*, 19, 2: 405–34.

Georgescu-Roegen, N. (1973) 'The entropy law and the economic problem' in H.E. Daly (ed.) *Towards a Steady State Economy* (San Francisco, CA: Freeman), 37–49.

Gilligan, C. (1983) *In a Different Voice: Psychological Theory and Women's Development* (Cambridge, MA: Harvard University Press).

Goldsmith, E., Allen, R., Allaby, M., Davoll, J. and Lawrence, S. (1972) *Blueprint for Survival* (Harmondsworth: Penguin).

Goodin, R. (1992) *Green Political Theory* (Cambridge: Polity Press).

Goodin, R. (1996) 'Enfranchising the earth, and its alternatives', *Political Studies*, 44, 5: 835–49.

Goodin, R. (2003) *Reflective Democracy* (Oxford: Oxford University Press).

Goodin, R. (2005) 'Deliberation Within', *Philosophy and Public Affairs*, 29, 1: 81–109.

Goodin, R. (2007) 'Enfranchising All Affected Interests and Its Alternatives', *Philosophy and Public Affairs*, 35, 1: 40–68.

Goodin, R. (2008) *Innovating Democracy: Democratic Theory and Practice after the Deliberative Turn* (Oxford: Oxford University Press).

Goodpaster, K. (1978) 'On Being Morally Considerable', *Journal of Philosophy*, 75: 308–25.

Gorz, A. (1982) *Fairwell to the Working Class: An Essay in Post-Industrial Socialism* (London: Pluto).

Gorz, A. (1985) *Paths to Paradise/On the Liberation From Work* (London: Pluto).

Gough, I. (2016) 'Welfare States and Environmental States: a comparative analysis', *Environmental Politics*, 14, 1: 47–67.

Gouldner, A. (1979) *The Future of Intellectuals and the Rise of the New Class* (Basingstoke: Macmillan).

Gray, J. (1993) *Beyond the New Right: Markets, Government and the Common Environment* (Cambridge: Cambridge University Press).

Griffen, S. (1978) *Woman and Nature: The Roaring Inside Her* (New York: Harper & Row).

Griscom, J. (1981) 'On Healing the Nature/History Split in Feminist Thought', *Heresies*, 13, 4: 4–9.

Gunnarsson, L. (2008) 'The Great Apes and the Severely Disabled: Moral Status and Thick Evaluative Concepts', *Ethical Theory and Moral Practice*, 11: 305–26.

Gutmann, A. and Thompson, D. (1996) *Democracy and Disagreement* (Cambridge, MA: Harvard University Press).

Gutmann, A. and Thompson, D. (2004) *Why Deliberative Democracy?* (Princeton, NJ: Princeton University Press).

Haas, P. (1990) *Saving the Mediterranean* (New York: Columbia University Press).

Habermas, J. (1996) *Between Fact and Norms* (Cambridge: Polity Press).

Hadley, J. (2015) *Animal Property Rights: A Theory of Habitat Rights for Wild Animals* (Lanham, MD: Lexington Books).

Hailwood, S. (2004) *How to be a Green Liberal* (Chesham: Acumen).

Hajer, M. (1997) *The Politics of Environmental Discourse: Ecological Modernization and the Policy Process* (Oxford: Clarendon Press).

Hansen, A. (ed.) (1993) *The Mass Media and Environmental Issues* (Leicester: Leicester University Press), 143–65.

Hardin, G. (1968) 'The Tragedy of the Commons', *Science*, 162: 1243–48.

Hardin, G. (1977) 'Living on a lifeboat' in G. Hardin and J. Baden (eds.) *Managing the Commons* (San Francisco, CA: Freeman).

Hart, H.L.A. (1967) 'Are there any Natural Rights?' in A. Quinton (ed.) *Political Philosophy* (Oxford: Oxford University Press), 53–66.

Harvey, E. (2016) 'What is behind the diesel cars emissions scandal?', *The Guardian*, 21 April.

Hay, P.R. (1988) 'Ecological values and western political traditions: From anarchism to fascism', *Politics* 8, 2: 22–29.

Hay, P. (2002) *A Companion to Environmental Thought* (Edinburgh: Edinburgh University Press).

Hayward, T. (1995) *Ecological Thought: An Introduction* (Cambridge: Polity).

Hayward, T. (1998) *Political Theory and Ecological Values* (Cambridge: Polity).

Hayward, T. (2005) *Constitutional Environmental Rights* (Oxford: Oxford University Press).

Hayward, T. (2006) 'Ecological Citizenship: Justice, Rights and the Virtue of Resourcefulness', *Environmental Politics*, 15, 3: 435–46.

Heilbroner, R. (1974) *An Inquiry into the Human Prospect* (New York: Norton).

Held, D. (2006) *Models of Democracy* (Cambridge: Polity Press, third edition).

Held, D. and McGrew, A. (1993) 'Globalization and the liberal democratic state', *Government and Opposition*, 2, 28: 261–88.

Hempel, L. (1996) *Environmental Governance: The Global Challenge* (Washington, DC: Island Press).

Heywood, A. (1992) *Political Ideologies: An Introduction* (Basingstoke: Macmillan).

Hirsch, F. (1977) *Social Limits to Growth* (London: Routledge and Kegan Paul).

Holland, B. (2014) *Allocating the Earth: A Distributional Framework for Protecting Capabilities in Environmental Law and Policy* (Oxford: Oxford University Press).

Homer-Dixon, T. (1991) 'On the threshold: environmental changes as causes of acute conflict', *International Security*, 16, 2: 76–116.

Homer-Dixon, T. (1999) *The Environment, Scarcity and Violence* (Princeton, NJ: Princeton University Press).

Horta, O. (2010) Debunking the idyllic view of natural processes: population dynamics and suffering in the wild, *Télos*, 17, 1: 73–88.

Huber, J. (1982) *Die verlorene Unschuld der Okologie* (Frankfurt am Main, Fischer Verlag).

Huby, M. (1998) *Social Policy and the Environment* (Buckingham: Open University Press).

Hunold, C. and Dryzek, J. (2002) 'Green Political Theory and the State: Context is Everything', *Global Environmental Politics*, 2, 3: 17–39.

Hurrell, A. (1995) 'International Political Theory and the Global Environment' in K. Booth and S. Smith (eds.) *International Relations Theory Today* (Cambridge: Polity Press).

Hurrell, A. and Kingsbury, B. (1992) 'The international politics of the environment: An introduction' in A. Hurrell and B. Kingsbury (eds.) *The International Politics of the Environment* (Oxford: Clarendon Press).

Hursthouse, R. (2000) *Ethics, Humans and Other Animals* (London: Routledge).

Inglehart, R. (1977) *The Silent Revolution: Changing Values and Political Styles Among Western Publics* (Princeton, NJ: Princeton University Press).

Inglehart, R. (1990) 'Values, ideology and cognitive mobilization in new social movements' in R. Dalton and M. Kuechler (eds.) *Challenging the Political Order: New Social and Political Movements in Western Democracies* (Cambridge: Polity Press), 43–66.

Inglehart, R. and Rabier, J. (1986) 'Political realignment in advanced industrial society: From class based politics to quality of life politics', *Government and Opposition*, 21, 4: 456–79.

IPCC (2007) Climate change 2007: Synthesis report. *Contribution of Working Groups I, II and III to the Fourth Assessment Report of the Intergovernmental Panel on Climate Change Geneva*, Switzerland.

IPCC (2014) Climate change 2014: Synthesis report. *Contribution of Working Groups I, II and III to the Fifth Assessment Report of the Intergovernmental Panel on Climate Change Geneva*, Switzerland.

Jackson, T. (2009) *Prosperity Without Growth: Economics for a Finite Planet* (London: Earthscan).

James, S. (2015) *Environmental Philosophy: An Introduction* (Cambridge: Polity).

Jamieson, D. (1998) 'Animal Liberation is an Environmental Ethic', *Environmental Values*, 7, 1: 41–57.

Janicke, M. (1985) *Preventive Environmental Policy as Ecological Modernisation and Structural Policy* (Berlin: Wissenschaftszentrum).

Keohane, R. (1989) *International Institutions and State Power* (Boulder, CO: Westview).

Keohane, R. and Nye, J. (1977) *Power and Interdependence: World Politics in Transition* (Cambridge, MA: Harvard University Press).

Kamminga, M. (2008) 'The Ethics of Climate Politics: four modes of moral discourse', *Environmental Politics*, 47, 4: 673–92.

Kant, I. (1965) *Metaphysics of Morals* (New York: Bobbs Merrill).

Kates, M. (2013) 'Justice, democracy and future generations', *Critical Review of International Social and Political Philosophy*, DOI 10.1080/13698230.2013.861655.

Kerschner, C. and O'Neil, D. (2015) 'Economic Growth and Sustainability' in H. Kopnina and E. Shoreman-Quinet (eds.) *Sustainability: Key Issues* (London: Routledge), 243–76.

Kheel, M. (1990) 'Ecofeminism and deep ecology: Reflections on Identity and Difference' in I. Diamond and G. Ornestein (eds.) *Reweaving the World: The Emergence of Ecofeminism* (San Francisco, CA: Sierra Club).

King, Y. (1981) 'Feminism and the revolt of Nature', *Heresies*, 13, 4: 12–16.

Kitsuse, J. and Spector, M. (1981) 'The labelling of social problems' in E. Rubington and M. S. Weinberg (eds.) *The Study of Social Problems* (New York: Oxford University Press), 198–206.

Kopnina, H. (2012) 'The Lorax Complex: Deep Ecology, Ecocentrism and Exclusion', *Journal of Integrative Environmental Sciences*, 9, 4: 235–54.

Kovel, J. (2007) *The Enemy of Nature: The End of Capitalism or the End of the World?* (London: Zed Books).

Kunzig, W. and Broecke, R. (2008) *Fixing Climate* (New York: Hill and Wang).

Kuper, R. (1997) 'Deliberating Waste: The Hertfordshire Citizens' Jury', *Local Environment*, 2, 2: 139–53.

Laferriere, E. and Stoett, P. (1999) *Ecological Thought and International Relations Theory* (London: Routledge).

Lafont, C. (2006) 'Is the Ideal of Deliberative Democracy Coherent?' in J. Besson and J. Marti (eds.) *Deliberative Democracy and its Discontents* (Aldershot: Ashgate), 3–25.

Latouche, S. (2009) *Farewell to Growth* (Cambridge: Polity Press).

Lauber, V. (1978) 'Ecology politics and liberal democracy', *Government and Opposition*, 13, 2: 199–217.

Layard, R. (2005) *Happiness: Lessons From a New Science* (New York: Penguin).
Lear, L. (1998) *Rachel Carson: Witness for Nature* (London: Penguin).
Leopold, A. (1949) *A Sand County Almanac* (Oxford: Oxford University Press).
Light, A. (2002) 'Restoring ecological citizenship' in B. Minteer and B. Pepperman Taylor (eds.) *Democracy and the Claims of Nature: Critical Perspectives for a New Century* (Lanham, MD, Boulder, CO: Rowman and Littlefield).
Linklater, A. (1996) 'The Achievements of Critical Theory' in S. Smith, K. Booth and M. Zalewski (eds.) *International Theory: Positivism and Beyond* (Cambridge: Cambridge University Press), 279–98.
Lively, J. (1975) *Democracy* (Oxford: Basil Blackwell).
Lomborg, B. (2001) *The Skeptical Environmentalist* (Cambridge: Cambridge University Press).
Lomborg, B. (2006) 'Climate change can wait. World health can't', *Observer*, 2 July 2006
Lomborg, B. (2007) *Cool it: The Skeptical Environmentalist's Guide to Global Warming* (New York: Alfred A. Knopf).
Lomborg, B. (ed.) (2010) *Smart Solutions to Climate Change* (Cambridge: Cambridge University Press).
Lomborg, B. and Rubin, O. (2002) 'Limits to Growth', *Foreign Policy*, 133: 42–44.
Lovelock, J. (2006) *The Revenge of Gaia* (London: Penguin).
Lowe, P. and Goyder, J. (1983) *Environmental Groups in Politics* (London: Allen & Unwin).
Lowe, P. and Rudig, W. (1986) 'Review article: Political ecology and the social sciences – The state of the art', *British Journal of Political Science*, 16: 513–50.
Luke, T. (2014) 'Karl Marx: Critique of Political Economy as Environmental Political Theory' in P. Cannavo, J. Lane and J. Barry (eds.) *Engaging Nature* (Cambridge, MA: MIT Press), 205–21.
Luskin, R., Fishkin, J. and Jowell, R. (2002) 'Considered opinions: Deliberative polling in Britain', *British Journal of Political Science*, 32, 3: 455–87.
Lyster, S. (1985) *International Wildlife Law* (Cambridge: Grotius).
Mackie, G. (2006) 'Does Democratic Deliberation Change Minds?', *Politics, Philosophy and Economics*, 5, 3: 279–304.
MacNeil, J., Winsemius, P. and Yakushiji, T. (1991) *Beyond Independence: The Meshing of the World's Economy and the Earth's Ecology* (Oxford: Oxford University Press).
Maddox, J. (1972) *The Doomsday Syndrome* (London: Maddox Editorial Ltd).
Malthus, T. (1968/1798) *An Essay on the Principles of Political Economy* (Harmondsworth: Penguin).
Martell, L. (1994) *Ecology and Society* (Cambridge: Polity).
Marx, K. and F. Engels (1848/1976) *The Communist Manifesto* (Harmondsworth: Penguin).
Maslow, A. (1954) *Motivation and Personality* (New York: Harper).
Mason, J. and Singer, P. (1990) *Animal Factories* (New York: Harmony, second edition).

Matthews, F. (1991) *The Ecological Self* (London: Routledge).
McCloskey, H. (1987) 'The Moral case for Experimentation on Animals', *Monist*, 70, 1: 64–82.
McCormick, J. (1989) *Acid Earth: The Global Threat of Acid Pollution* (London: Earthscan).
McCormick J. (1992) *The Global Environmental Movement* (London: Belhaven).
McMahan, J. (2002) *The Ethics of Killing: Problems at the Margins of Life* (Oxford: Oxford University Press).
Meadowcroft, J. (2005) 'From Welfare State to Ecostate' in J. Barry and R. Eckersley (eds.) *The State and the Global Ecological Crisis* (Cambridge, MA: MIT Press), 3–23.
Meadows, D.H., Meadows, D.L., Randers, J. and Behrens III, W. (1972) *The Limits to Growth: A Report for the Club of Rome's Project on the Predicament of Mankind* (New York: Universe).
Meadows, D.H., Meadows, D.L. and Randers, J. (1992) *Beyond the Limits: Global Collapse or a Sustainable Society: Sequel to the Limits to Growth* (London: Earthscan).
Meadows, D., Randers, J. and Meadows, D. (2005) *The Limits to Growth: The Thirty Year Update* (London: Earthscan).
Mellor, M. (1992) 'Ecofeminist, ecofeminine or ecomasculine', *Environmental Politics*, 1, 2: 29–51.
Mellor, M. (1997) *Feminism and Ecology* (Cambridge: Polity).
Mellor, M. (2006) 'Socialism' in A. Dobson and R. Eckersley (eds.) *Political Theory and the Ecological Challenge* (Cambridge: Cambridge University Press), 35–50.
Mellos, K. (1988) *Perspectives on Ecology: A Critical Essay* (New York: St Martins).
Merchant, C. (1980) *The Death of Nature: Women, Ecology and the Scientific Revolution* (New York: Harper and Row).
Meyer, J. (2011) 'We Have Never Been Liberal: the environmentalist turn to liberalism and the possibilities for social criticism', *Environmental Politics*, 20, 3: 356–73.
Middleton, N. (2013) *The Global Casino: An Introduction to Environmental Issues* (fifth edition) (London: Routledge).
Mill, J.S. (1909) *Principles of Political Economy* (London: Longmans).
Miller, M. (1995) *The Third World in Global Environmental Politics* (Buckingham: Open University Press).
Mills, M. (1996) 'Green Democracy: The search for an ethical solution' in B. Doherty and M. de Geuss (eds.) *Democracy and Green Political Thought* (London: Routledge), 96–114.
Mills, M. and King, F. (2000) 'Ecological constitutionalism and the limits of deliberation and representation' in M. Saward (ed.) *Democratic Innovation: Deliberation, Representation and Association* (London: Routledge).
Milton, K. (2002) *Loving Nature: Towards an Ecology of Emotion* (Abingdon: Routledge).
Mishan, E. (1967) *The Costs of Economic Growth* (London: Penguin).

Mol and Sonnenfeld (2000) 'Ecological Modernisation around the World: Perspectives and Critical Debates', *Environmental Politics*, 9, 1.

Monbiot, G. (1999) 'Poisoning those who are poor', *The Guardian*, 20 May.

Morgenthau, H.J. (1962) *Politics in the Twentieth Century* (Chicago, IL: University of Chicago Press).

Mouffe, C. (2005) *On the Political* (London: Routledge).

Mutz, D. (2008) 'Is Deliberative Democracy a Falsifiable Theory?', *Annual Review of Political Science*, 11: 521–38.

Myers, N. (1989) 'The future of forests' in L. Friday, L. and R. Laskey (eds.) *The Fragile Environment* (Cambridge: Cambridge University Press), 22–40.

Myers, N. and Simon, J. (1994) *Scarcity or Abundance: A Debate on the* (New York: Norton).

Nagel, T. (2005) 'The Problem of Global Justice', *Philosophy and Public Affairs*, 33: 113–47.

Naess, A. (1973) 'The shallow and the deep, long range ecology movement. A summary', *Inquiry*, 16: 95–100.

Newell, P. (2006) 'Environmental justice movements: Taking stock, moving forward', *Environmental Politics*, 15, 4: 656–60.

Nibert, D. (2002) *Animal Rights/Human Rights: Entanglements of Oppression and Liberation* (Washington, DC: Rowman and Littlefield).

Niemeyer, S. (2004) 'Deliberation in the Wilderness: Displacing Symbolic Politics', *Environmental Politics*, 13, 2: 347–72.

Nordhaus, W. and Boyer, J. (2000) *Warming the World: Economic Models of Global Warming* (Cambridge, MA: The MIT Press).

Nordlinger, E. (1967) *The Working Class Tories: Authority, Deference and Stable Democracy* (London: MacGibbon & Kee).

North, R. (1995) *Life on a Modern Planet: A Manifesto for Progress* (Manchester: Manchester University Press).

Norton, B. (1991) *Toward Unity Among Environmentalists* (Oxford: Oxford University Press).

Norton, B. (1999) 'Ecology and opportunity: intergenerational equity and sustainable options' in A. Dobson (ed.) *Fairness and Futurity: Essays on Environmental Sustainability and Social Justice* (Oxford: Oxford University Press), 118–50.

Norton, P. (2003) 'A Critique of Generative Class Theories of Environmentalism and the Labour-Environmentalist Relationship', *Environmental Politics*, 12, 4: 96–119.

Nozick, R. (1974) *Anarchy, State and Utopia* (Oxford: Basil Blackwell).

Nussbaum, M. (2004) 'Beyond Compassion and Humanity: Justice for Nonhuman Animals' in C. Sunstein and M. Nussbaum (eds.) *Animal Rights: Current Debates and New Directions'* (Oxford: Oxford University Press), 299–320.

Nussbaum, M. (2006) *Frontiers of Justice: Disability, Nationality, Species Membership* (Cambridge, MA: Harvard University Press).

Nussbaum, M. (2011) *Creating Capabilities: The Human Development Approach* (Cambridge, MA: Harvard University Press).

O'Neill, J. (1993) *Ecology, Policy and Politics: Human Well-Being and the Natural World* (London: Routledge).

O'Neill, J. (2003) 'Varieties of Intrinsic Value' in A. Light and Holmes Rolston III (eds.) *Environmental Ethics: An Anthology* (Oxford: Blackwell), 131–42.

O'Neil, K. (2009) *The Environment and International Relations* (Cambridge: Cambridge University Press).

Ophuls, W. (1973) 'Leviathan or Oblivion?' in H. Daly (ed.) *Toward a Steady State Economy* (San Francisco: Freeman), 215–30.

Ophuls, W. (1977) *Ecology and the Politics of Scarcity* (San Francisco: W.H. Freeman & Co).

O'Riordan, T. (1976) *Environmentalism* (London: Pion).

Owen, D. (2012) 'Constituting the polity, constituting the demos: On the place of the all affected interests principle in democratic theory and in resolving the democratic boundary problem', *Ethics and Global Politics*, 5, 3: 129–52.

Paehlke, R. (1989) *Environmentalism and the Future of Progressive Politics* (New Haven, CT: Yale University Press).

Palmer, C. (2010) *Animal Ethics in Context* (New York: Columbia University Press).

Page, E. (2006) *Climate Change, Justice and Future Generations* (Cheltenham: Edward Elgar).

Page, E. (2008) 'Distributing the burdens of climate change', *Environmental Politics*, 17, 4: 556–75.

Parfit, D. (1984) *Reasons and Persons* (Oxford: Oxford University Press).

Parkinson, J. (2006) Deliberating in the Real World (Oxford: Oxford University Press).

Parsons, H. (1978) *Marx and Engels on Ecology* (Westport, CT: Greenwood).

Passmore, J. (1980) *Man's Responsibility for Nature* London: Duckworth, second edition).

Pateman, C. (1970) *Participation and Democratic Theory* (Cambridge: Cambridge University Press).

Paterson, M. (1999) 'Globalisation, Ecology, and Resistance', *New Political Economy*, 4, 1: 129–146.

Paterson, M. (2000) *Understanding Global Environmental Politics* (Basingstoke: Palgrave Macmillan).

Paterson, M. (2006) 'Theoretical perspectives on international environmental politics' in M. Betsill, K. Hochstetler and D. Stevis (eds.) *Palgrave Advances in International Environmental Politics* (Basingstoke: Palgrave Macmillan), 54–81.

Paterson, M. (2007) 'Environmental Politics: Sustainability and the Politics of Transformation', *International Political Science Review*, 28, 5: 545–56.

Paterson, M. (2016) 'Political economy of the greening of the state' in T. Gabrielson, C. Hall, J. Meyer and D. Schlosberg (eds.) *The Oxford Handbook of Environmental Political Theory* (Oxford: Oxford University Press), 475–90.

Pearce, D. et al. (1993) *Blueprint 3: Measuring Sustainable Development* (London: Earthscan).

Pellow, D. (2004) *Garbage Wars: The Struggle for Environmental Justice in Chicago* (Cambridge, MA: MIT Press).

Pepper, D. (1986) *The Roots of Modern Environmentalism* (London: Routledge).

Pepper, D. (1993) *Eco-Socialism: From Deep Ecology to Social Justice* (London: Routledge).

Pepper, D. (1993a) 'Anthropocentrism, humanism and eco-socialism: A blueprint for the survival of ecological politics', *Environmental Politics* 2, 3: 428–52.

Peterson, M. and Sandin, P. (2013) 'The Last Man Argument Revisited', *Journal of Value Inquiry*, 47: 121–33.

Petts, J. (2001) 'Evaluating the Effectiveness of Deliberative Processes: Waste Management Case Studies', *Journal of Environmental Planning and Management*, 44, 2: 207–26.

Pickering, J. and Barry, C. (2012) 'On the Concept of Climate Debt: Its Moral and Political Value', *Critical Review of International Social and Political Philosophy*, 15, 5: 667–85.

Plumwood, V. (1993) *Feminism and the Mastery of Nature* (London: Routledge).

Plumwood, V. (2006) 'Feminism' in A. Dobson, and R. Eckersley (eds.) *Political Theory and the Ecological Challenge* (Cambridge: Cambridge University Press).

Pogge, T. (2002) *World Poverty and Human Rights: Cosmopolitan Responsibilities and Reforms.* (Cambridge: Polity Press).

Porritt, J. (1984) *Seeing Green* (Oxford: Basil Blackwell).

Porritt, J. and Winner, D. (1988) *The Coming of the Greens* (London: Fontana).

Posner, E. and Weisbach, D. (2010) *Climate Change Justice* (Princeton, NJ: Princeton University Press).

Pratt, V. (2000) *Environment and Philosophy* (London: Routledge).

Putnam, R. (2001) *Bowling Alone: The Collapse and Revival of American Community* (New York: Simon & Schuster)

Rachels, J. (1990) *Created From Animals: The Moral Implications of Darwinism* (Oxford: Oxford University Press).

Rawls, J. (1971) *A Theory of Justice* (Oxford: Oxford University Press).

Ray, E. and Parson, S. (2016) 'Reimagining Radical Environmentalism' in T. Gabrielson, C. Hall, J. Meyer and D. Schlosberg (eds.) *The Oxford Handbook of Environmental Political Theory* (Oxford: Oxford University Press), 580–92.

Raz, J. (1986) *The Morality of Freedom* (Oxford: Clarendon Press).

Redclift, M. (1984) *Development and the Environmental Crisis: Red or Green?* (London: Methuen).

Regan, T. (1980) 'Utilitarianism, Vegetarianism and Animal Rights', *Philosophy and Public Affairs*, 9, 4: 305–24.

Regan, T. (1984) *The Case for Animal Rights* (London: Routledge).

Regan, T. (1991) *The Thee Generation: Reflections on the Coming Revolution* (Philadelphia, PA: Temple University Press).

Regenstein, L. (1985) 'Animal rights, endangered species and human survival' in P. Singer (ed.) *In Defence of Animals* (Oxford: Blackwells).

Richardson, D. (1997) 'The politics of sustainable development' in S. Baker, M. Kousis, D. Richardson and S. Young (eds.) *The Politics of Sustainable Development: Theory, Policy and Practice Within the European Union* (London: Routledge), 43–60.

Ritvo, H. (1987) *The Animal Estate* (Cambridge, MA: Harvard University Press).

Rockstrom, J., Steffen, W., Noone, K. et al. (2009) 'A Safe Operating Space for Humanity', *Nature*, 461: 472–75.

Rodman, J. (1983) 'Four forms of ecological consciousness reconsidered' in D. Scherer and T. Attig (eds.) *Ethics and the Environment* (Englewood Cliffs, NJ: Prentice-Hall), 82–92.

Rolston, H. (1975) 'Is There an Ecological Ethic?', *Ethics*, 85, 1: 93–109.

Rosenau, J. (1993) 'Environmental Challenges in a Turbulent World' in K. Conca and R. Lipschutz (eds.) *The State and Social Power in Global Environmental Politics* (New York: Columbia University Press).

Routley, R. (1993) 'Is There a Need for a New, an Environmental, Ethic?' in M. Zimmerman et al. (eds.) *Environmental Philosophy: From Animal Rights to Radical Ecology* (New Jersey: Prentice Hall), 12–21.

Rowell, A. (1996) *Green Backlash: Global Subversion of the Environmental Movement* (London: Routledge).

Rowlands, M. (1998) *Animal Rights: A Philosophical Defence* (Basingstoke: Macmillan).

Rowlands, M. (2009) *Animal Rights: Moral Theory and Practice* (Basingstoke: Palgrave Macmillan).

Ryfe, D. (2005) 'Does Deliberative Democracy Work?', *Annual Review of Political Science*, 8: 49–71.

Ryle, M. (1988) *Ecology and Socialism* (London: Century Hutchinson).

Sagoff, M. (1988) *The Economy of the Earth: Philosophy, Law and the Environment* (Cambridge: Cambridge University Press).

Sale, K. (1980) *Human Scale* (New York: Coward, Cann and Geoghegan).

Sale, K. (1984) 'Bioregionalism-a new way to treat the land', *The Ecologist*, 14: 167–73.

Sandbach, F. (1978) 'The Rise and Fall of the Limits to Growth Debate', *Social Studies of Science*, 8, 4: 495–520.

Sandel, M. (1998) *Liberalism and the Limits of Justice* (Cambridge: Cambridge University Press, second edition).

Saunders, B. (2011) 'Defining the Demos', *Politics, Philosophy and Economics*, 11, 3: 280–301.

Saunders, F. (2015) 'Planetary Boundaries: at the threshold ... again: sustainable development ideas and politics', *Environment Development and Sustainability*, 17: 823–35.

Saurin, J. (1993) 'Global Environmental Degradation, Modernity and Environmental Knowledge', *Environmental Politics*, 2, 4: 46–64.

Saward, M. (1993) 'Green theory', *Environmental Politics*, 2, 3: 509–12.

Saward, M. (1996) 'Must democrats be environmentalists?' in B. Doherty and M. de Geus (eds.) *Democracy and Green Political Thought: Sustainability, Rights and Citizenship* (London: Routledge), 79–96.

Saward, M. (2001) 'Reconstructing Democracy: Current Thinking and New Directions', *Government and Opposition*, 36, 4: 559–81.

Schlosberg, D. (2004) 'Reconceiving Environmental Justice: Global Movements and Political theories', *Environmental Politics*, 13, 3: 517–40.

Schlosberg, D. (2007) *Defining Environmental Justice: Theories, Movements and Nature* (Oxford: Oxford University Press).

Schlosberg, D. (2012) 'Climate Justice and Capabilities: a Framework for Adaptation Policy', *Ethics and International Affairs*, 26, 4: 445–61.

Schlosberg, D. (2013) 'Theorising Environmental Justice: the expanding sphere of a discourse', *Environmental Politics*, 22, 1: 37–55.

Schoon, N. (1990) 'Acid rain that earns Britain a black mark', *The Independent*, 26 March 1990.

Schumacher, E.F. (1973) *Small is Beautiful: Economics as if People Mattered* (London: Blond and Briggs).

Scruton, R. (2006) 'Conservatism' in A. Dobson and R. Eckersley (eds.) *Political Theory and the Ecological Challenge* (Cambridge: Cambridge University Press), 7–19.

Seager, J. (1993) *Earth Follies: Coming to Feminist Terms with the Global Environmental Crisis* (New York: Routledge).

Sen, A. (1993) 'Capabilities and Well-Being' in M. Nussbaum and A. Sen (eds.) *The Quality of Life* (Oxford: Clarendon Press), 30–53.

Sen, A. (2009) *The Idea of Justice* (Cambridge, MA: Harvard University Press).

Shearman, D. and Wayne-Smith, J. (2007) *The Climate Change Challenge and the Future of Democracy* (Westport, CT: Praeger).

Simon, J. (1981) *The Ultimate Resource* (Princeton, NJ: Princeton University Press).

Simon, J. and Kahn, H. (eds.) (1984) *The Resourceful Earth: A Response to Global 2000* (New York: Basil Blackwell).

Simpson, S. (1990) *The Times Guide to the Environment* (London: Times Books).

Singer, P. (1983) *The Expanding Circle: Ethics and Sociobiology* (Oxford: Oxford University Press).

Singer, P. (1990) *Animal Liberation* (London: Cape).

Singer, P. (1985) 'Famine, Affluence, Morality' in C. Beitz (ed.) *International Ethics* (Princeton, NJ: Princeton University Press).

Singer, P. (2002) *One World: The Ethics of Globalization* (Melbourne: Text Publishing).

Singer, F. and Avery, D. (2007) *Unstoppable Global Warming: Every 1500 Years* (Lanham, MD: Rowman and Littlefield).

Smith, M. (1998) *Ecologism: Towards Ecological Citizenship* (Buckingham: Open University Press).

Smith, G. (2003) *Deliberative Democracy and the Environment* (London: Routledge).

Smith, K. (2016) 'Environmental Political Theory, Environmental Ethics, and Political Science: Bridging the Gap' in T. Gabrielson, C. Hall, J. Meyer and D. Schlosberg (eds.) *The Oxford Handbook of Environmental Political Theory* (Oxford: Oxford University Press), 105–15.

Stavins, R. (2000) 'Market-Based Environmental Policies' in P. Portney and R. Stavins (eds.) *Public Policies for Environmental Protection* (Washington, DC: Resources for the Future, second edition).

Steinbock, B. (1978) 'Speciesism and the Idea of Equality', *Philosophy*, 53: 247–56.

Steiner, H. (1998) 'Working Rights' in M. Kramer, N. Simmonds and H. Steiner (eds.) *A Debate Over Rights: Philosophical Enquiries* (Oxford: Clarendon Press), 233–301.

Steiner, J., Bachtiger, A., Sporndli, M. and Steenbergen, M. (eds.) (2004) *Deliberative Politics in Action. Analysing Parliamentary Discourse* (Cambridge: Cambridge University Press).

Stephens, P. (2001) 'The green only blooms amid the Millian flowers: a reply to Marcel Wissenburg', *Environmental Politics*, 10, 3: 43–47.

Sterba, J. (1998) *Justice for Here and Now* (Cambridge: Cambridge University Press).

Stern Report (2009) www.occ.gov.uk/activities/stern.htm

Stone, C. (1974) *Should Trees Have Standing: Toward Legal Rights for Natural Objects* (Los Altos, CA: Kaufmann).

Sunstein, C. (2008) 'The Law of Group Polarization' in J. Fishkin and P. Laslett (eds.) *Debating Deliberative Democracy* (Chichester: John Wiley & Sons), 80–101.

Swatuk, L. (2006) 'Environmental Security' in M. Betsill, K. Hochstetler and D. Stevis (eds.) *Palgrave Advances in International Environmental Politics* (Basingstoke: Palgrave Macmillan), 203–36.

Swift, A. (2014) *Political Philosophy* (Cambridge: Polity).

Swyngedouw, E. (2009) 'The Antinomies of the post-political city. In search of a democratic politics of environmental production', *International Journal of Urban and Regional Research*, 33, 3: 601–20.

Taylor, D. (1999) 'Wage slaves, throw off your chains!', *The Guardian*, 23 June.

Taylor, P. (1986) *Respect For Nature: A Theory of Environmental Ethics* (Princeton, NJ: Princeton University Press).

Taylor, N. and Twine, R. (ed.) (2014) *The Rise of Critical Animal Studies: From the Margins to the Centre* (London: Routledge).

Thomas, C. (1992) 'The United Nations Conference on Environment and Development of 1992 in context', *Environmental Politics*, 4, 1: 250–66.

Thompson, D. (2010) 'Representing future generations: Political presentism and democratic trusteeship', *Critical Review of International Social and Political Philosophy*, 13, 1: 17–37.

Tranter, B. (1996), 'The Social Bases of Environmentalism in Australia', *Australian and New Zealand Journal of Sociology*, 32, 2: 63–85.

Tremmel, J. (2006) 'Establishing intergenerational justice in national constitutions' in J. Tremmel (ed.) *The Handbook of Intergenerational Justice* (London: Edward Elgar).

Tuxill, J. (1999) 'Appreciating the benefits of plant biodiversity' in L. Brown et al. (eds.) *State of the World 1999* (New York: Norton), 96–114.

Vanderheiden, S. (2008) *Atmospheric Justice: A Political Theory of Climate Change* (Oxford: Oxford University Press).

Vanderheiden, S. (2016) 'Environmental and Climate Justice' in T. Gabrielson, C. Hall, J. Meyer and D. Schlosberg (eds.) *The Oxford Handbook of Environmental Political Theory* (Oxford: Oxford University Press), 321–32.

Victor, P. (2008) *Managing Without Growth: Slower by Design, Not Disaster* (Cheltenham: Edward Elgar).

Vincent, A. (1993) 'The character of ecology', *Environmental Politics*, 2, 2: 248–76.

Vincent, A. (1995) *Modern Political Ideologies* (second edition) (Oxford: Blackwell).

Vincent, A. (1998) 'Is Environmental Justice a Misnomer?' in D. Boucher and P. Kelly (eds.) *Social Justice from Hume to Walzer* (London: Routledge), 120–40.

Vogler, J. (2005) 'In Defence of International Environmental Cooperation' in J. Barry and R. Eckersley (eds.) *The State and the Global Ecological Crisis* (Cambridge, MA: MIT Press), 229–53.

Vogler, J. (2008) 'Environmental Issues' in J. Baylis, S. Smith and P. Owens (eds.) *The Globalisation of World Politics* (Oxford: Oxford University Press).

Waltz, K. (1979) *Theory of International Politics* (Reading, MA: Addison-Wesley).

Walzer, M. (1999) 'Deliberation and What Else?' in S. Macedo (ed.) *Deliberative Politics: Essays on Democracy and Disagreement* (Oxford: Oxford University Press), 58–69.

Ward, B. and Dubos, R. (1972) *Only One Earth: The Care and Maintenance of a Small Planet* (London: Andre Deutsch).

Warren. K. (ed.) (1994) *Ecological Feminism* (London: Routledge).

Warren, W. (2006) 'Book Review', *Society and Natural Resources*, 19: 377–80.

Weale, A. (1992) *The New Politics Of Pollution* (Manchester: Manchester University Press).

Weale, A. (2007) *Democracy* (Basingstoke: Palgrave, second edition).

Wendt, A. (1999) *Social Theory of International Politics* (Cambridge: Cambridge University Press).

Wenz, P. (1988) *Environmental Justice* (Albany, NY: State University of New York Press).

Weston, J. (ed.) (1986) *Red and Green: The New Politics of the Environment* (London: Pluto Press).

Wilkinson, R. and Pickett, K. (2009) *The Spirit Level: why more equal societies almost always do better* (London: Penguin).

Williams, R. (1986) 'Hesitations before socialism', *New Socialist*, September: 34–36.

Williams, M. (1996) 'International political economy and global environmental change' in J. Vogler and M. Imber (eds.) *The Environment and International Relations* (London: Routledge), 41–58.

Wilson, H. (2016) 'Edmund Burke: The Nature of Politics' in T. Gabrielson, C. Hall, J. Meyer and D. Schlosberg (eds.) *The Oxford Handbook of Environmental Political Theory* (Oxford: Oxford University Press), 153–71.

Wissenburg, M. (1993) 'The idea of nature and the nature of distributive justice' in A. Dobson and P. Lucardie (eds.) *The Politics of Nature: Explorations in Green Political Thought* (London: Routledge), 3–20.

Wissenburg, M. (1998) *Green Liberalism* (London: UCL Press).

Wissenburg, M. (2006) 'Liberalism' in A. Dobson and R. Eckersley (eds.) *Political Theory and the Ecological Challenge* (Cambridge: Cambridge University Press), 20–34.

Worcester, R. (1997) 'Public opinion and the environment' in M. Jacobs *Greening the Millennium: The New Politics of the Environment* (Oxford: Blackwell), 160–73.

World Commission on Environment and Development (1987) *Our Common Future* (Oxford: Oxford University Press).

Wurzel, R. (2012) 'The Ecological Challenge to nation States: From Limits to Growth to Ecological Modernisation' in J. Connelly and J. Hayward (eds.) *The Withering of the Welfare State* (London: Palgrave Macmillan).

Wurzel, R. and Connelly, J. (eds.) *The European Union as a Leader in International Climate Change Politics* (London: Routledge).

Young, J. (1990) *Post Environmentalism* (London: Belhaven).

Yearley, S. (1992) *The Green Case* (London: Routledge).

Young, I. (1990) *Justice and the Politics of Difference* (Princeton, NJ: Princeton University Press).

Young, I. (2000) *Inclusion and Democracy* (Oxford: Oxford University Press).

Young, S. (1992) 'The different dimensions of green politics', *Environmental Politics*, 1, 1: 9–44.

Young, S. (1993) *The Politics of the Environment* (Manchester: Baseline Books).

INDEX

www.ingramcontent.com/pod-product-compliance
Ingram Content Group UK Ltd.
Pitfield, Milton Keynes, MK11 3LW, UK
UKHW021034150926
475561UK00003B/103

9 781137 607393